M000096746

Guided BY THE Faith of Christ

Seeking to Stop
Violence and Scapegoating

Stephen R. Kaufman, M.D.

Vegetarian Advocates Press
Cleveland, Ohio

Vegetarian Advocates Press
Cleveland, Ohio

Second printing, October 2008

ISBN 9-7809716-67648

Contents

Acknowledgements

I am deeply indebted to the scholarship and original contributions of many insightful thinkers, particularly René Girard (1923–) and Ernest Becker (1924–1974), and others, including James Alison, Gil Bailie, Paul Nuechterlein, Norman O. Brown, Ron Liefer, David Loy, and Daniel Liechty. In addition I owe an intellectual debt to the stimulating comments by contributors to the René Girard (girard.topic@ecunet.org) and Ernest Becker (ernestbecker-l@listserv.ilstu.edu) Internet discussion lists.

I would like to thank the following people who offered helpful feedback on this book: Rev. Linda McDaniel, David Sztybel, Norm Phelps, Rev. Frank L. Hoffman, Mary T. Hoffman, Rev. Lisa Hadler, Tom Youngjohn, George Matejka, Thomas Suits, Gregory Massey, Richard Hershey, Betsy Wosko, Roberta Kalechofsky, Lorena Mucke, Bruce Friedrich, Annika Spalde, Denise Williams, Stephen Augustine, Gracia Fay Ellwood, Vasu Murti, Anthony Marr, Rev. Stephen Adams, Pelle Strindlund, and Nathan Braun, and others whom I have not named here. I give special thanks to my wife Betsy for helpful comments and support, and her father Thomas Suits, a retired classics professor, for assistance in translating passages from the original Greek Bible. Finally, I am grateful for the excellent editorial assistance of Elaine Payne. Of course, none of these people are responsible for any theological conclusions or inferences I have drawn.

More people than is possible to name have assisted my ongoing spiritual journey. I would like to highlight Rev. Dr. James Antal, the former pastor of Plymouth Church of Shaker Heights, UCC, who baptized me. I also acknowledge Rev. Frank Hoffman, Nathan Braun, Stephen Webb, and Maynard Clark for their roles in creating and developing the Christian Vegetarian Association (CVA) of which I am chair. Though my study of violence has spanned several decades, my focus has been narrowed more recently as I have learned, with shock and dismay, about the massive, pervasive, and unnecessary abuse of animals in the United States. I have found my participation in the CVA's ministry, made possible by the dedicated efforts of many people, important and meaningful. I dedicate my work with the CVA, as I seek to dedicate my life, to God.

Introduction

Violence has plagued humanity since the beginning of human history. Understanding the causes of violence is one of the most pressing concerns of our time, given that violence is a leading cause of death and destruction. In the last century, there were genocides in Germany, Rwanda, Cambodia, and elsewhere, and well over 100 million people died in wars. In global terms, the proliferation of weapons of mass destruction now threatens human civilization and the rest of God's creation. In personal terms, perhaps in part because I have Jewish roots, the Holocaust haunts me. While the Holocaust's scope was distinctive, episodes of similar brutality have occurred throughout human history.

The task of understanding and diminishing violence challenges science, religion, and ethics. Attempts to address violence scientifically are crucial, because it is likely that averting human and worldwide disaster will require our understanding the psychological, sociological, and anthropological underpinnings of violence. Most people aim to be kind, compassionate, and charitable. Yet many people will, under the right circumstances, engage in acts that strike onlookers as irrational and cruel. If science cannot explain this paradox, it is unlikely that science will do much to prevent the potentially catastrophic consequences of violence.

In attempting to explain why humans so often participate in violence, several thinkers have provided important insights. This book will borrow heavily from the works of René Girard, Ernest Becker, and scholars who have been influenced by them. Their contributions to psychology, sociology, and cultural anthropology have aided our understanding of human violence.

Can Christianity assist in addressing the problem of violence? Some, noting the many past episodes of violence in the name of Jesus and God, have regarded Christianity as part of the problem, not part of the solution. In trying to discern Christianity's teachings about violence, it is important to explore God's and Jesus' relationship with humanity and the rest of creation. I will focus on the Bible in the hopes that this time-honored text provides insight and inspiration. Do reasonable biblical interpretations accord with science's discoveries about humanity, including reasons for our propensity for violence? Does the Bible provide reasonable explanations for why there is so much suffering among humans and other animals (hereafter "animals")? Can the Bible teach us how to live peacefully with each other?

The Bible has many themes, including God's plan throughout history, how to form community, and how to lead a meaningful life. Violence has important implications for these topics, and the Bible is replete with stories of killing, from the murder of Abel to the Hebrews' bloody conquest of the Promised Land to the murders of John the Baptist, Jesus, and St. Stephen. Does the Bible teach that violence is an inevitable consequence of struggling to survive outside Eden, or does the Bible offer guidance for surmounting violence?

Though many biblical stories depict God promoting peace and compassion, there are also many accounts in which God seems to endorse violence and, occasionally, to participate in violence. It has been tempting for people of many religious faiths to regard their god as wrathful and as hating the same people whom they hate. I will examine biblical passages some Christians have used to portray God as wrathful and violent, and I will suggest different ways to regard these passages—ways that would condemn the Crusades, the Inquisition, the anti-Semitic pogroms, the enslavement of humans, the widespread mistreatment of animals, and the other harmful things that Christians and other people have done.

I do not claim that this book's framework, heavily informed by Becker and Girard, fully accounts for the rich Judeo-Christian tradition and the many themes it addresses. I contend that, with this framework in mind, many of the Bible's stories and passages encourage us to conclude that God cares about all creation, that our human tendency to participate in violence undermines God's desires, and that our adopting the faith of Christ[1] (Chapter 6) offers a path toward overcoming scapegoating. I will not attempt to offer a complete systematic theology, nor will I try to fully review those systematic theologies that offer different understandings of Christian Scripture and faith. Though I cannot discuss all biblical texts, I will try to identify crucial biblical stories and passages that relate to this book's theses. Similarly, although I cannot address all possible objections to my views, I will attempt to anticipate readers' questions and comments, particularly when my perspectives seem to conflict with currently popular theologies.

The first chapter explores certain insights from psychology, sociology, and cultural anthropology that have helped us to understand why humans have so much trouble coexisting peacefully. In particular, I will argue that scapegoating, which involves attributing excess guilt to victims, derives from powerful human psychological needs. In addition to its injustice, scapegoating undermines the biblical ideal of peaceful coexistence among all creatures. Can Christian faith address those needs in a nonviolent, just manner? Chapters 2 through 11 relate the scapegoating process to a range of biblical stories and themes. In particular, I will consider how an anthropological approach to the Bible can enhance our understanding and application of the accounts in the Hebrew Scriptures; the Ten Commandments; the stories about Jesus' life, death, and resurrection; and Jesus' and Paul's teachings. I will also explore the roles of faith, love, forgiveness, healing, peacemaking, and prophecy in overcoming the human tendency to participate in violence and scapegoating. Chapter 12 critically evaluates certain theological frameworks and parts of the Bible that many contemporary Christians have used to justify "sacred" violence done in God's name. Chapter 13 applies the book's biblical and anthropological frameworks to some contentious contemporary social issues.

This book has three principal, related themes. First, I aim to show that the faith of Christ, which Christians are called to emulate, is centrally about acting with love, compassion, and mercy. Second, the faith of Christ has both individual and communal elements. Though we often have the sense that we act independently, our social networks play crucial roles in molding the values, beliefs, and attitudes that direct our behavior. This, I think, is why the Bible emphasizes community as the vehicle for righteousness. Third, I hold that adopting the faith of Christ does not require "believing six impossible things before breakfast."[2] This faith has firm foundations in insights from the social sciences, from our own personal experiences, and from the inspired writers of the time-honored Scriptures that Christians revere. Truth, as best we can discern it, should guide our faith, because Jesus said, "For this I was born, and for this I have come into the world, to bear witness to the truth" (John 18:37[3]), and "If you continue in my word, you are truly my disciples, and you will know the truth, and the truth will make you free" (John 8:31–32). I will attempt to show that the faith of Christ as depicted by the Bible relates to profound truths that can lead to personal and communal peace, harmony, and salvation.

Chapter 1: The Scapegoating Process

The sacrificial process prevents the spread of violence by keeping vengeance in check. . . . The more critical the situation, the more "precious" the sacrificial victim must be.
René Girard, *Violence and the Sacred*, 1979, p. 18.

Man is . . . split in two: he has an awareness of his own splendid uniqueness in that he sticks out of nature with a towering majesty, and yet he goes back into the ground . . . in order blindly and dumbly to rot and disappear forever. It is a terrifying dilemma to be in and to have to live with.
Ernest Becker, *The Denial of Death*, 1973, p. 26.

Overview

Self-esteem helps assuage our innate fear of death. Often, we gain self-esteem by succeeding in competitions for objects of desire. The reason we compete is that mimesis largely directs desire; in other words, we tend to formulate our desires by regarding what other people seem to want. Such "acquisitive mimetic desires" readily lead to rivalries and conflicts that can damage relationships and split communities. The human "solution" to this problem has been to find one or more scapegoats whom people blame for larger communal discord. Many, and perhaps all, primal religions have transformed scapegoating into "sacrifice" and regarded it as "sacred" and ordained by the gods. In order for scapegoating to restore communal peace and order, people must not realize that they have attributed excessive guilt to the victims. Consequently, people have always found it difficult to identify their own scapegoating, including today.

The Fear of Death

Ernest Becker, standing on the shoulders of such intellectual giants as Otto Rank and Norman O. Brown, asserted that fear of death profoundly shapes human psychology and sociology. In *The Denial of Death*,[1] Becker noted that we humans are animals, and indeed there is voluminous fossil, anatomic, physiologic, and behavioral evidence that humans and animals share a common ancestry. Behavioral studies, Becker observed, have shown that humans and animals fear dangers that might result in death. However, Becker asserted that we are different from animals in recognizing that we are always vulnerable to death and in knowing that mortality is inevitable. People who have survived events in which death appeared imminent, such as an airplane losing control, often describe a feeling of terror, and "higher" animals can probably feel something analogous. In essence, death casts a terrifying shadow over our lives, because we know that, like passengers on a doomed airplane, our lives are headed toward death.[2] It is not specifically death of the body that terrifies the human psyche. Becker maintained that, at some level of consciousness, we are terrified of the extinction of the self—the "I" that forms our sense of personal identity. How do we maintain equanimity in the face of impending death?

Rather than feel terror constantly, which would leave us paralyzed with fear, people employ psychological defense mechanisms. One such defense mechanism involves projecting our fears about things that we cannot control onto things that we can manage and perhaps control. In *The Right Stuff*,[3] Tom Wolfe describes military test pilots, who were often killed in unavoidable crashes, frequently driving their cars at high, dangerous speeds. Evidently, they transferred their fear of death in flight, an event over which they likely felt little control, to fear of dying in an automobile accident, over which they had more control and which therefore was a more manageable fear. Evidently, driving dangerously gave them a sense of mastery over death, which made them feel more confident that they could survive hazardous flying missions.

Another defense mechanism involves projecting fears onto other people. Children often have fears related to their vulnerability, which they may project onto a fat or clumsy child, whom they regard as a manifestation of what it means to be vulnerable. The children then tease or otherwise humiliate that child, giving themselves a sense of mastery over their fears. Likewise, a middle-aged man with fears of bodily decay might project those fears onto his similarly aging wife, and then despise her for manifesting such bodily decay. He might then feel justified in having an affair with a younger woman, upon whom he projects his yearning for everlasting youth. Entire communities can project fears of vulnerability or death onto a single person, whom they treat with contempt or violence.

Self-Esteem: A Fundamental Desire

We like to think that we are captains of our own ship—that we make decisions based on sound, reality-based judgments. However, we may not realize that our conscious or unconscious hopes and fears can greatly distort our values and judgments. Some fears relate to events in early childhood that we cannot recall, and some fears are so terrifying that our minds try to repress them from consciousness. An example of the latter, Becker maintained, is fear of death, and he asserted that self-esteem helps to reduce or repress mortality fears. The need for self-esteem might be most obvious in children, who feel vulnerable because they are surrounded by bigger and stronger people and because many accidental injuries come as complete surprises to them. They need self-esteem to have enough confidence to explore the world. Children gain a sense of self-esteem by mastering tasks, and they frequently mimic their parents and older siblings who, from the child's perspective, are confident and competent.

Adults similarly find that self-esteem helps diminish fears that they will be killed.[4] Those with good self-esteem tend to believe that they have the skills needed to manage dangerous situations and avoid harm. Also, self-esteem engenders a feeling of importance, which leads to a sense of "it couldn't happen to me." Although this might appear irrational to outside observers, many people feel that, if they are important, God—or one of God's agents—will protect them from harm.

Self-esteem helps provide a sense of immortality. There is an intense psychological desire to believe that the "self"—the sense of individual identity that has a personal name and that we sense does not fundamentally change throughout our lives—will not perish when the body ceases to function. We invariably generate mental images of what the world will be like after we die. If our self-esteem is low, we will more likely expect people to quickly forget us. If we feel important, we will more likely envision people long remembering us. These latter images will help give us a sense of what Becker called "death transcendence," which can comfort us psychologically. Another way that self-esteem can promote a sense of death transcendence is that we can believe that, as "good" people, God will reward us with everlasting paradise.

We are not born with the knowledge of what we must accomplish to warrant the status of a "good person," and throughout our lives we receive a multitude of messages from our parents, friends, community, and larger culture about what we should do. Men typically learn that they should be self-reliant, tough, physically and emotionally stoical, and physically attractive, and they should also be modest, helpful, and sensitive. Women should be cooperative, empathetic, attentive to the needs of their children, and they should also be independent and able to control their emotions, and they should not waste their intellectual and other talents. Our families, friends, and wider culture also teach men and women that good people also have a "good" or wealthy family, a heterosexual orientation, and an absence of major disabilities or deformities—features over which we have no control. With so many difficult and often conflicting cultural standards, gaining and maintaining a strong sense of self-esteem is a Herculean task. Consequently, I think the quest for self-esteem contributes significantly to widespread feelings of inadequacy, frustration, and anger in our culture and in many other cultures. Having trouble identifying the source of these feelings, we tend to blame family, friends, or others for our sense of malaise, which strains bonds within our families and communities.

Self-Esteem and Culture

Becker argued that one's culture is the vehicle through which one gains self-esteem.[5] We are not born knowing whether or not we are "good" or "admirable," and we generally judge our worth according to the assessments of those in our community. Consequently, "image" is very important to most people, but what generates a positive image?

What it means to be "good" varies widely between cultures. Some cultures admire aggressiveness, and others value nonviolence. In some cultures, people revere those who acquire the most; in others those who share the most gain the highest regard. Another way that culture promotes self-esteem is by providing stories that indicate that one's own group is admirable. This helps explains why people tend to divide themselves into smaller groups defined by such features as geographic location, nation, ethnicity, socioeconomic class, or gender. They then regard "their" group as better than "other" groups.

If we only gain self-esteem by being good in relation to others, the quest for self-esteem invariably leads to competition. We can compete at an individual level, such as in sporting contests or the pursuit of higher social standing, or we can compete at a communal level, perhaps by trying to assert the superiority of our own group over other groups.

Similar to humans, when animals compete for desired objects, such as food or a mate, they exhibit intense emotions. Afterward, however, animals' emotions quickly subside; one animal has won and the other has lost, and they tend to change their focus to other concerns. Generally, they appear show little or no resentment. While many animals seek to improve their social status, and those with inferior rank may challenge those with higher rank at opportune times, animals seem to have much less interest in revenge than humans. For humans, getting the desired object means far more than satisfying an immediate desire—self-esteem is often at issue. Losing can hurt our self-esteem and make us feel more mortal. Consequently, losing can engender long-standing anger, bitterness, and resentment.

Importantly, self-esteem is not something we either have or do not have, like some material object. Our sense of self-esteem might increase or diminish throughout life, depending on our success in the marketplace, at sports, and in other competitions, and whether the larger culture confirms our "success" by praise, awards, and other means of recognition. Though we struggle continuously to maximize self-esteem, we can never have enough self-esteem to fully quell death-fears. Furthermore, if we feel humiliated, our self-esteem can suffer as long as the offender goes unpunished for the offense. The desire for revenge can so trouble our minds that we might be unable to feel at peace.

Because self-esteem is tied to the satisfaction of desires, we often have the sense that we "need" things that actually are not essential for survival, or even for contented living. In terms of self-esteem, what matters most is not what we own, but what we own in relation to others. Consequently, people generally want to have more, regardless of how much they have. Even those with the most will not feel satisfied, because they can never fully eliminate their fears related to death.

Support for Becker's Theories

It seems that Becker's theories about fear of death strike some people as obviously true, though others remain doubtful. In an effort to test Becker's theories, researchers have developed Terror Management Theory (TMT). A broad range of TMT-inspired experiments have shown that, as Becker's theories would predict, thinking about death encourages people to defend their culture, to prefer people of similar ethnic background, and even to become violent.[6,7] In many of these experiments, researchers asked people to think about what it would feel like to die and then to be dead, a concept the researches called "mortality salience." Interestingly, nearly all denied that the mental images were upsetting. But subsequent testing indicated that they were deeply troubled by the images, suggesting that people often do not recognize how much death disturbs them. Many people find Becker's theories unconvincing because they do not perceive a fear of death within themselves. However, experimental findings, including those based on TMT, have indicated that people repress their mortality fears.

TMT has also shown the psychological importance of self-esteem. One series of experiments involved altering subjects' self-esteem by telling them that, after a bogus test, they either did or did not have good personality traits. Those subjects who were told they had desirable personality traits were subsequently less troubled when they thought about their own mortality compared to those who were given a negative assessment after taking the bogus test.

Further, studies have indicated that for many people raising self-esteem is a buffer against the effects of mortality salience on their opinions about their culture. For example, some American subjects took a bogus "personality test" and were told that they had positive attributes, and they were then instructed to think about their dying and being dead. These subjects were not nearly as offended by an essay that criticized American culture as other subjects who were similarly exposed to mortality salience but were told that they had only a neutral score on the bogus personality test.[8]

Remarkably, TMT studies have also demonstrated that mortality salience increases aggression. In one study, subjects were asked to read a political essay with which they agreed or disagreed. They were then told that they were to administer a taste test to the essay's writer, in which the subjects supplied a quantity of hot sauce for the writer to taste. Those subjects exposed to mortality salience gave far greater amounts of hot sauce to writers of essays with which they disagreed than writers with whom they agreed. Subjects not exposed to mortality salience gave only slightly more hot sauce to writers with whom they disagreed than writers with whom they agreed, but the difference was not statistically significant.[9]

Other evidence that self-esteem is a primary human concern comes from daily experience. The intense anger many of us feel upon being humiliated appears to relate to our hurt sense of self-esteem. Similarly, self-esteem appears to underlie fierce competition in sports, business, and many other walks of life. We may even gain self-esteem vicariously—consider how good many of us feel when "our" sports team wins.

Becker described our "inner-newsreel," the ongoing story in our minds about our lives that includes recollections of our past successes and triumphs and our past failures and humiliations.[10] Becker wrote,

The scenario of self-value is not an idle film hobby. The basic question the person wants to ask and answer is "Who am I?" "What is the meaning of my life?" and "What value does it have?" And we can only get answers by reviewing our relationships to others, what we do to others and for others, and what kind of responses we get from them. Self-esteem depends on our social role, and our inner-newsreel is always packed with faces—it is rarely a nature documentary.[11]

The inner newsreel is an ongoing story in which we, as directors, tend to try to convince ourselves, as viewers, that we are important and good. If we have suffered a recent setback, Becker asserted, we tend to try to balance this unpleasant occurrence by conjuring up memories of related, past events in which we fared better. If someone verbally humiliates us and we cannot think of an effective retort, we tend to reflect later on what we should have said, or we reflect on past episodes in which we effectively "put down" someone else. Similarly, we try to balance our shortcomings by reminding ourselves that we belong to a local, religious, ethnic, national, or other group that is good. When we sleep, the inner newsreel keeps running, but we are unable to consciously add favorable images to counter the unpleasant ones. Consequently, our dreams often consist of nightmarish feelings of shame, dependency, vulnerability, and worthlessness.

Carol Tavris and Elliot Aronson have studied how we tend to view our actions and the actions of others in ways that confirm our self-worth. When people were asked to relate an episode in which they were hurt or angered and an episode in which they hurt or angered someone else, people described similar events, such as betrayals or acts of unkindness. When they had been the perpetrators, they nearly always claimed that they were justified by circumstances. Further, they generally described the effects of their hurtful actions as temporary and not severe. In contrast, when they had been victims, they usually considered the motive to be inexplicable or immoral, and they regarded the harmful effects on themselves as severe and long-lasting.[12]

Our minds frequently rewrite history to generate stories that confirm our goodness and value. One psychologist had been convinced that his mother had tried to dissuade him from accepting admission and a scholarship at Stanford, because Stanford was far from home. Twenty-five years later, when reading old letters he had written to his mother, he was shocked to read how his mother passionately urged him to accept Stanford's offer over that of a local university. Evidently, his mind had rewritten the story. He recalled the conflict with his mother, but his mind had changed the facts such that he believed he was the one with the personal strength to leave his family and the wisdom to pursue the attractive offer from Stanford.[13]

Importantly, fear of death and pursuit of self-esteem do not determine human behavior. Rather, they influence the choices we make, often in ways that we have trouble identifying. I think that Becker's theories help explain behaviors that otherwise sometimes seem irrational and inexplicable, including how people frequently fear and hate members of different cultures and how people often have intense desires for vengeance against those who have humiliated them. One of the principal ways that we gain self-esteem is by procuring objects of desire, but what generates desires?

Mimetic Desire

Humans naturally imitate each other and adopt each other's mannerisms, opinions, and attitudes. Imitation is crucial to human social development, because young children instinctively and unconsciously learn social skills and language by mimicking other people's behaviors. Similarly, adults, by observing others' attitudes and behaviors, recognize a wide range of threats and opportunities. However, such imitation can also be the vehicle for spreading rumors, stigmas, and paranoia. New ideas, some productive and desirable and others harmful and destructive, can spread like highly contagious diseases throughout a community.

One of René Girard's seminal insights was that human desires derive from imitation, and he called this process "mimesis." All people have innate desires, including desires for food, being touched, and having social interactions. When there are no choices, people desire whatever meets their basic biological needs. During famines, people desire any food, not specific foods. Most often, there are myriad ways that we can satisfy our desires, and we are not born knowing which we should choose. Girard, borrowing from psychology, sociology, cultural anthropology, and literature, concluded that people determine what *they* want to acquire by seeing what *other* people seem to value. This "acquisitive mimetic desire," Girard noted, is generally unconscious.

To illustrate acquisitive mimetic desire, a child in a room filled with toys will frequently want the one toy with which another child is playing. Although adults generally deny that their own desires are mimetic, the advertising industry exploits the powerful pull of mimetic desire with actors or celebrities who portray goods or services as extremely desirable. What passes for desirable can be quite arbitrary, can change over time, and can even vary among different subcultures. This shows how acquisitive desire is mimetic and often has little relation to the actual qualities of the object of desire. In some cultures, women have sought men who can boast of accomplishments on the battlefield, while in other cultures women have found intellectual or sensitive men most attractive. Similarly, during the Renaissance many men regarded a full figure as a standard of beauty among women, while popular female models and movie stars today generally have more slender figures.

We often have little conscious awareness of mimetic desires as they grow within us. We like to think that we determine our own desires and that we choose objects for their inherent desirability. Such thinking helps maintain our sense of self-esteem. If we are attracted to a certain person, we will generally convince ourselves that this is because that person is inherently attractive, not because other people find that person attractive or because that person meets certain culturally defined standards of attractiveness. No doubt, conscious rationality often plays an important role in our discernments; we often draw on our life experiences about which choices give us comfort and pleasure. However, I think Girard has correctly identified mimesis as an important factor in many of our specific choices.

Acquisitive Mimetic Desire

Humans cannot avoid mimetic desire—it is part of our innate make-up. Importantly, there are different kinds of mimetic desire. *Acquisitive* mimetic desire, in which we want to acquire what other people have, invariably leads to mimetic rivalries and, often, conflicts. Girard identified another kind of mimetic desire that he called "good mimesis," in which our model is someone with whom we cannot fall into rivalry. As we will see, Jesus was such a model.

One problem with acquisitive mimetic desire, which appears to be the predominant kind of mimetic desire among humans, is that it invariably leads to scarcity. Acquisitive mimetic desire encourages people to seek the same things, and demand invariably outstrips supply. Scarce things can include certain foods, material goods, attractive mates, or esteem by peers (because not everyone can be highly regarded). Indeed, scarcity often makes things more desirable, because obtaining scarce things demonstrates one's success, which increases self-esteem. However, failure to satisfy our acquisitive mimetic desires generates resentfulness toward those who have gained the desired items.

In any culture in which self-esteem is associated with wealth, such as ours, there must be some individuals who most people regard as "poor." What it means to be poor varies among societies, because wealth and poverty are relative terms. A wealthy person in one community might be regarded poor in a much wealthier community. Because self-esteem is tied to wealth in materialistic cultures, those with wealth tend to walk a fine line. They often want people to envy them, because this validates their sense of self-esteem. However, wealthy people do not want the envy to lead to such marked resentment and anger that poorer people will feel entitled to steal from or even kill wealthy people. Indeed, whenever acquisitive mimetic desires principally motivate people, humiliations abound and outbreaks of violence loom.

Anger

When we feel humiliated or when our self-esteem is wounded we often feel angry. Because attributing our anger to our own shortcomings would further erode self-esteem, we tend to direct the anger outward. When angry, we often have difficulty recognizing our contribution to our dissatisfaction or interpersonal conflict. Instead, we often think of reasons why the target of our anger deserves our disregard or contempt. Even though we like to see ourselves as rational and objective, a strong emotion such as anger can readily trump reason. Indeed, as Freud and many others have recognized, intense feelings, hopes, fears, and fantasies—of which we are often unaware or only dimly aware—profoundly influence our interpretations of stories, symbols (such as the flag), and experiences (such as sexual intercourse). Children tend to be good observers and poor interpreters; adults are often the same way, particularly when a situation arouses powerful emotions or when self-justification is at issue. Although we generally have little difficulty identifying when other people seem self-deluded, most people seem convinced that their own conclusions are consistently well-grounded in fact and logic.

Nevertheless, anger undermines our reasoning abilities. If I hurt someone physically or emotionally in anger, I would probably recognize my anger, but I would probably not attribute my violence to that anger. My inner newsreel would not declare, "I hurt him because I was angry." I would likely consider my violence an appropriate response to his offense. I might regret that my anger had overpowered my self-control, but in defense of my self-esteem, I would likely conclude that his behavior, not my angry feelings, precipitated the violence.

Anger can rupture relationships, which can be very destructive to families or small communities. Our being mimetic creations can augment the destructive consequences of anger, because we often respond mimetically to others' emotional displays. Consequently, if I am angry at someone, that person will often respond with anger, which can increase my own angry feelings. As our anger escalates, anger's ability to override reason has important implications. One of those implications, crucial for Girardian thought, is that anger can be easily displaced from the original object (e.g., my boss who has belittled me) to a substituting object (e.g., somebody weaker than I am, such as an underling at work, a family member, or an animal).

Communities can similarly displace anger. In primal cultures (i.e., those without modern technology), survival has often depended on cooperative hunting and cooperative protection from predators. How can such cultures maintain cohesiveness, given the human tendency to develop mimetic rivalries that threaten to destroy bonds of loyalty or even lead to violence? The solution has often been to find a scapegoat. If everyone can agree that one individual is responsible for the growing hard feelings that threaten a community, then killing or expelling that individual can unify the community and defuse the conflict.

Scapegoating and Communal Discord

Scapegoating is effective at relieving the angry feelings and resentments that invariably arise as a consequence of mimetic rivalries. In order for scapegoating to restore tranquility, those involved in scapegoating must believe that the scapegoat is truly responsible for the social disharmony. The scapegoat victim may not agree, but the voice of the victim has little relevance, because the scapegoat is ostracized, exiled, or killed. It is possible for the accusation against the scapegoat to be unanimous, because reason easily becomes anger's servant. As anger spreads mimetically among the people, they readily convince each other that the scapegoat is responsible for the problem. Just as desire is mimetic, the accusation is mimetic. To outside observers, the collective accusation may seem totally unfounded and irrational. To those caught up in mimetic accusation, their minds clouded by anger and their accusation reinforced by their neighbors, the accusation seems obviously true.

Who are the scapegoats? In general, scapegoats have been peripheral members of a community who can be abused without much fear of retaliation by family or friends. Indeed, what makes them different can be what people come to see as the manifestation of their evilness or contemptibility. Scapegoats often have distinctive physical or psychological attributes, such as an unusual facial feature, a limp, or an inclination toward psychotic delusions.

It is not hard to think of examples of scapegoating. Children often bond by collectively teasing and humiliating one child, who might be overweight, have bad teeth, or have some other physically distinctive feature. Many families have a "black sheep" who always makes trouble at family gatherings. If it were not for their collective contempt at the black sheep's behavior, they would likely find themselves at odds with each other. At a societal level, examples of communal violence—the Nazi Holocaust, the "ethnic cleansing" in the Balkans, the genocide in Rwanda, the Salem witch hunt—have a common theme of scapegoating one individual or group of individuals in order to restore communal peace and a sense of well-being.

Guilt, Shame, and the Scapegoating Process

For purposes of this book, I will regard scapegoating as the process by which guilt or shame is transferred from the responsible party to another individual or group. Whenever we fail to succeed, we feel less adequate, and when others criticize or condemn us, our sense of shame substantially increases. Such feelings can significantly injure our sense of self-esteem, and there is a strong temptation to modify the content of our inner newsreel from personal inadequacy to someone else deserving blame for our recent failure, social mishap, or general low social standing. When we fail as individuals, we are tempted to blame spouses for not supporting us, bosses for not recognizing our assets, or promotion policies that we consider unfair. When we fail as communities, we might blame social, political, or military leaders, or we might blame a minority group. The collective humiliation that the German people felt as a result of the loss of World War I and the Treaty of Versailles encouraged them to seek a scapegoat. Consequently, Hitler's virulent claim that the Jews had betrayed the German people and were responsible for Germany's defeat appealed to many Germans. What about natural disasters, such as a drought, insect infestation, epidemic, or earthquake? To the degree that people regard such disasters as judgments, they will tend to search for someone to blame.

It is crucial that those involved in scapegoating convince themselves that the scapegoat is blameworthy.[14] Several things make it easier to believe the falsehood about the scapegoat's guilt. First, when communities scapegoat, there is mimetic accord regarding the victim's guilt. Second, the community often behaves in ways that hide the truth. For example, when Stephen was stoned, people found it necessary to avoid hearing the innocent victim. They "cried out with a loud voice and stopped their ears" (Acts 7:57). Third, since everyone has faults and has behaved badly at some time, anyone can be accused of misdeeds that deserve banishment or punishment. Therefore, in general, the injustice of scapegoating is not that the victims are blameless, though sometimes they are; it is that the victims are not as blameworthy as the scapegoaters claim. Fourth, scapegoaters often believe that their actions accord with divine will, citing sacred texts or pronouncements from religious leaders for validation. In these situations, people generally regard the scapegoating as a "sacred" activity and see it as "sacrifice." Not all scapegoating is sacrificial, but all or nearly all manifestations of sacrificial violence involve scapegoating.

One individual can scapegoat another. When we have conflicts with other people, we often have good reasons to be angry about an offense against us. However, almost always our own statements or actions have added to the hostility. Consequently, when we accuse someone else of being responsible for the conflict, we almost always shift our part of the blame for the conflict onto them, and this is scapegoating.

Not all harm and violence is scapegoating. There are times when people consciously harm other individuals to achieve material gain or elevated social standing, or to avoid the consequences of their own bad decisions or behavior. Such conscious selfishness reflects a rejection of God's love and goodness, and, if depravity were the overwhelming motivation for human destructiveness, there would seem to be little hope for a better future. However, I think many actions that appear to represent mere selfishness primarily relate to a desire to relieve deep-seated feelings of humiliation and guilt. There is a largely unconscious shift in blame. Indeed, if the shift in blame were conscious, it would not relieve the perpetrator's sense of guilt and shame: If I knew I was blaming someone else for my own misdeeds or shortcomings, I would not feel more worthy.

The reason for regarding scapegoating in terms of the motivations of the perpetrators is both practical and moral. From a practical standpoint, there seems to be little hope to appeal to the hearts and minds of those who consciously choose to cause harm. In contrast, those who harm innocent individuals but genuinely believe that their actions are righteous and just could, with further insight and understanding, mend their ways. From a moral standpoint, unlike those who intentionally, malevolently harm others, those who do not recognize their scapegoating do not deserve condemnation. For those who "know not what they do" (Luke 23:34), we should still love the sinner while we hate the sin.

Scapegoating in Primal Communities

Girard has argued that scapegoating has been central to the development of culture. In order to make such a claim, he looked at the religious myths of primal cultures. Why not look at our own culture? First, our culture has experienced the growth of sciences that have raised doubts about religious myths. Our sciences have provided evidence that the creation myths of all religions, if taken literally, are not reasonable. Second, our culture has been influenced by nonsacrificial, nonscapegoating teachings, including those of Jesus, in ways that make it difficult for people to believe that scapegoating violence really resolves social crises. We will explore this in more detail later in the book. Third, it is difficult to provide good examples of what people in our culture will agree constitutes scapegoating: people do not recognize their own scapegoating, because the act of scapegoating must be hidden in order for it to work.

In primal cultures, scapegoating has frequently occurred in response to communal crises such as natural disasters or hostilities related to mimetic rivalries. Typically, people start to suspect a marginal member of the community of casting evil spells or of angering the gods by violating a sacred taboo, such as against blasphemy or incest. With strong communal desires to blame someone for the crisis, the accusation becomes increasingly universal because accusation is mimetic. The community, convinced by each other of the scapegoat's guilt, typically kills or exiles the scapegoat. Remarkably, the crisis often ends after the scapegoating, which appears to confirm that the scapegoat victim was indeed responsible. Previously feuding communal members become united in their hatred of and collective actions against the scapegoat. In the case of a natural disaster, killing the scapegoat has often seemed to resolve the problem. Earthquakes and most other natural disasters rarely recur in the immediate future; droughts tend to end on their own accord; and epidemics usually run their course. Consequently, events seem to confirm the validity of the collective accusation. Because scapegoating often accompanies resolution of the crisis and because the resolution has often seemed miraculous, people have tended to regard their communal violence as divinely ordained. These miraculous events encourage the development of myths in which people come to believe not only that sacred violence ends crises but that such violence can also prevent crises. This is the connection between scapegoating and sacrificial violence.

Myth, Ritual, and Taboo

According to Girard, primal communities have developed religions that feature myths, rituals, and taboos in an attempt to avoid future crises. The myths have described how their gods have wanted sacrifices. Indeed, the "truth" of these myths has seemed obvious precisely because "sacred," sacrificial violence has generally preceded resolution of a crisis. Also, subsequent sacrifices have seemed to maintain peace and tranquility, which appears to prove that sacrifices have pleased the divine. Some myths have included instructions for sacrificial rituals. These rituals have aimed to reenact both the circumstances leading to the crisis and the original sacrificial violence that relieved the crisis. I refer readers to Girard's *Violence and the Sacred* for numerous examples.[15]

Myths also identify taboos, and children in primal communities learn from the elders that violating taboos will anger the gods and create chaos and destruction. In general, taboo objects and activities are those that have seemed to contribute to fierce mimetic rivalries. Many cultures have taboos that delineate the power, wealth, and marriage opportunities of the members of different castes, classes, genders, or other social groups. By restricting social opportunities, fewer people compete for power, luxuries, and attractive mates. In ancient Egyptian society, one's choices were severely restricted by one's class. There was virtually no personal freedom, and there was also little mimetic rivalry.[16]

Because people have been taught that taboos derive from instructions by the gods, and because people have intuitively understood that taboos help maintain social order, they have generally defended taboos vigorously. Contemporary social reformers often contend that taboos have served primarily to maintain oppression of women, minorities, and poor people, and therefore they regard taboos as sinister attempts to selfishly exclude some people from the community's bounty. To a degree, this is likely true. However, there have also been widespread fears that violating taboos would lead to chaos. Frequently, many members of a community believe that social order would break down if members of a traditionally subservient group were put in a dominant position. This helps explain the vigor with which many Southern whites once defended the subservient position of blacks. Prior to passage of civil rights legislation in the 1960s, the notion of a black foreman telling a white laborer what to do was abhorrent to many whites. Whites often said, "Everyone gets along down here." However, "getting along" often included the threat of a violent response to blacks who "didn't know their place." Thankfully, America has made great strides toward eradicating many unjust taboos, though many people continue to experience the pernicious effects of racism, sexism, classism, and other forms of prejudice.

The vast majority of Americans now applaud efforts to eliminate unjust laws and social arrangements that have persisted by use of violence or threat of violence. However, removing unjust barriers to freedom of choice can increase the number of people seeking the same objects of desire, and it can increase the number of people who feel damaged self-esteem from failing to obtain these objects of desire. Consequently, a potential cost of social justice can be greater mimetic rivalry, conflict, and resentment. In later chapters, we will consider ways to increase social justice without increasing the risks of social unrest and violence.

Despite myths, rituals, and taboos, social crises still occur in primal cultures. Sometimes, after several generations, people forget how terrifying the social crisis was, and more people violate the taboos that have helped maintain social order. Alternatively, droughts, epidemics, or other natural disasters can cause people to question whether their ritual sacrifices are adequate to appease their gods. In both situations, there exists what Girard has called a "sacrificial crisis," in which people lose faith in existing myths, rituals, and taboos; social hierarchies break down; and mimetic rivalries lead to an "all-against-all" environment of chaos and destructiveness. Terrified, people seek an explanation for the discord, and typically rumors spread that one or more people have violated a sacred taboo or have cast evil spells. Once the accusation is unanimous or nearly so (those with doubts know to keep quiet, lest they be accused), victims are collectively killed or banished.

Sacrifice and the Scapegoating Process

Whenever there has been a sacrificial crisis, a new scapegoat has been found, and myths, rituals, and taboos have been created or modified. Girard reviewed the anthropological literature, and he found that, *all* primal cultures either engage in blood sacrifice or have myths and rituals that relate back to blood sacrifices. Further, Girard has found that all primal cultures have myths that recall one or more killings that were central to the creation of the world.[17] This has led Girard to conclude that "sacred," collective, scapegoating violence originally brought people together. Further, ritual sacrifice, as an offshoot of scapegoating, has continued to unify communities ever since. Many primal peoples attribute past or current sacrificial crises to angry gods, and they perform sacrifices to appease those gods.

Girard has asserted that the social, political, and religious hierarchies that are central to human cultures derive from the scapegoating process.[18] But he has not argued that there has been a single event in the remote past for each culture. Rather, cultures develop over long periods of time. As long as a culture's myths, rituals, and taboos seem to prevent a sacrificial crisis, few people will challenge them. New "knowledge" arises when, in response to a sacrificial crisis, communities scapegoat one or more victims and develop new myths, rituals, and taboos that, they believe, will prevent future sacrificial crises.

Contrary to what many critics of religion have maintained, Girard holds that religions attempt to control violence, but they typically do not address the underlying cause of violence. According to Girard, mimetic rivalries have generated violence, and these rivalries have always threatened to tear communities apart. Religious sacrifices have been attempts to control violence by substituting small doses of sacred violence for widespread outbreaks of profane violence.

While I have found compelling Girard's contention that collective scapegoating has been important in binding primal communities together, other factors have likely also played important roles, such as the need to avoid predators and to seek prey. Likewise, the scapegoating process has probably helped mold myths, rituals, taboos, and other components of human culture, but so have other concerns, such as the search for cosmic understanding and meaning among self-conscious creatures. Similarly, Girard has asserted that religious sacrifices have often reduced the amount of human violence, but religions have also encouraged people to exterminate entire tribes, such as occurred during the Hebrew conquest of the Promised Land.

Girard has argued that accusation is mimetic, and that the scapegoating victims are often quite arbitrary. Therefore, to the degree that the scapegoating process is the foundation of culture, I think it is reasonable to posit that culture has been generated by scapegoating animals as well as people, because people can easily accuse animals of being vehicles of evil spirits. I wonder whether primal communities have found animals particularly attractive as scapegoating victims, because small communities would often find animals easier to accuse, abuse, and replace than people. Indeed, the term "scapegoat" derives from the ritual described in Leviticus 16:21-22, in which the community collectively transferred its guilt onto a goat, and the goat was then sent into the wilderness, where the goat would likely die. Further, as we will see, animals are often victims of scapegoating today. If scapegoating animals as well as humans has played a crucial role in generating our culture, we must find ways to stop scapegoating both animals and humans if we are to establish and maintain a just and peaceful society.[19]

Evidence for the Scapegoating Process

Girard and subsequent Girardians have looked at a wide range of primal myths and found that they consistently both reveal and conceal the scapegoating process. They reveal the scapegoating process in that their creation stories typically relate how killing or expelling an individual created order from chaos. The myths conceal the scapegoating process by asserting that the gods approved the killing or expulsion. Those responsible for scapegoating are the ones who describe the event with mythic stories, and they believe that the victims were guilty and deserved their fate.

The central Hindu creation myth describes Purusha as a primal human being with grotesque features, a symbol for chaos. The gods dismembered Purusha, and everything on earth derives from his body. The priestly caste comes from Purusha's head, the noble-warrior caste from his arms, the populace from his thighs, and the untouchables from his feet. If one deconstructs the myth anthropologically, one may regard Purusha as representing a real person who was collectively murdered, bringing peace and order to a chaotic community.[20] Similarly, the Nawatl Aztec's ritual for the renewal of fire, recorded circa 1500, recreated the communal killing of a victim as the origin of culture. First, the people burned all their blankets and destroyed their pottery, which reenacted an original, precivilized state. Then, they placed a sacrificial victim atop a pyramid, and the priest cut his chest cavity open and ripped out his heart. Afterwards, the priest placed a bowl of tinder in the chest cavity and started a fire by rubbing sticks. The new fire lit a torch that subsequently lit torches for everyone else in the community.[21] Evidently, this ritual symbolized regeneration of the community by virtue of sacred violence ordained by their god. Girard's book *The Scapegoat*[22] deconstructs many myths in an effort to discern the historic truth behind the myth.

In a given culture, there is no reason to doubt the truthfulness of the myths. The myths offer an explanation for the origins of the universe in general and the given culture in particular. Further, the myths describe how people should lead their lives and to what achievements they should aspire. The myths satisfy the intense human desire to understand where we come from, why we are here, and where we are going. However, if the scapegoating process generates human culture, then falsehoods about the victim's guilt underlie the myths that tell people what is right and wrong, what is meaningful and irrelevant, and what is true and false about the mysterious universe in which they live. A Girardian reading of the Bible suggests that the falsehood that underpins the scapegoating process is what "has been hidden since the foundation of the world" (Matthew 13:35). For Christians, Jesus exposed this falsehood, though other religious leaders have also demonstrated insight. Around the eighth Century BCE, the latter Hebrew prophets denounced the unjust violence of blood sacrifices.[23] In the sixth century BCE, the Buddha and Pythagoras rejected animal sacrifices, and the Buddha also condemned the caste system.[24]

Contemporary Scapegoating: Humans and Animals

Most people believe that in this modern, "enlightened" era we have abandoned sacrificial violence. However, as in primal cultures, we likely have trouble identifying our own scapegoating. Cultures tend to defend, often at the expense of personal liberties or justice, such nebulous concepts as national security, family, civilization, and religion. Those who stand up and criticize communally sanctioned violence might be accused of being anti-family or unpatriotic and become themselves ostracized, harassed, or imprisoned.

The rise of humanism has helped secure human rights and, evidently, helped reduce the scapegoating of humans. However, it is possible that our culture participates in scapegoating as much as any previous culture, except that the victims today are much more commonly animals rather than humans. A clue that violence against animals has a "sacred" scapegoating element is that animal defenders often arouse great anger from the animal-exploiting general public. To be sure, one reason is that prohibiting activities that harm animals threatens to impact people's lifestyles, because many enjoy the taste of animal flesh, the "glamour" of animal furs, the challenge of the hunt, the thrill of the bullfight, the competition of the rodeo, and the supposed benefits of animal experimentation. However, given that there is no serious threat to nearly all of the ongoing, widespread uses of animals, the hostility toward animal advocates suggests that there is a sacred element to activities that involve harming animals. If so, attempts to restrict those activities might raise doubts about the validity of certain myths, rituals, and taboos with which people orient their lives.

Many animal protectionists note that humans are among the members of the animal kingdom, a view that seems to contradict the myth[25] that humans are special creations with unique importance to God or the universe. Humanity's continuity with the natural world in general and animals in particular reminds people of their own mortality. To many people, animals seem to lead meaningless lives characterized by struggle and ending in anonymous death. The rich and meaningful lives that animals actually have are not always evident to the casual observer.[26] When people kill animals, it gives them a sense of superiority and a sense that they are fundamentally different from the animals. The act of eating animals—consuming their very bodies—can reinforce this conviction.[27] This might help explain the horror many people feel when hearing stories about people being eaten by animals. It reminds people that they are vulnerable to death and also that they are made of flesh. Along these lines, many cultures deal with corpses in ways that help people avoid seeing human flesh decompose, and the prohibition against cannibalism is nearly universal.

Many of us feel guilty about our socially unacceptable desires. For many of us, a sense that God condemns anyone who has such illicit desires augments these guilty feelings. It seems that many people resolve this concern by projecting their own forbidden sexual, violent, or other desires onto animals.

Regarding sexuality, human sexuality has myriad mimetic, cultural, and biological influences that frequently conflict with each other, and these often lead to inner turmoil and interpersonal difficulties. It can be tempting for humans to deny their own "immoral" sexual desires—nearly everyone has desires that would violate one or more of our culture's many taboos related to sexuality. Though animals are generally very selective about their sexual partners, on the surface animals often seem uninhibited about sex, perhaps because many animals, unlike people, have sexual intercourse in the presence of other members of their species. Consequently, people can deny their unacceptable sexual desires by feeling contempt for animals, who seem to engage in unrestrained sexual behavior.

Regarding violence, people have similarly tried to justify their own violence by distinguishing their violence—which they regard as righteousness or justice—from that of animals. This is ironic, because it appears that humans are far more inclined than animals to seek vengeance, and animals rarely kill or seriously injure each other when fighting over food, territory, or sexual mates. Indeed, people often call sexual predators or violent criminals "animals" or "beasts," which implicitly denies the possibility that the rest of us might have similar sexual or violent desires. Because we desperately want to deny our own guilt and shame, and because we want to believe that we are good and worthy of God's favor, there are strong motivations to project our own unwanted desires onto scapegoats. Animals, who cannot protest unfair characterizations, can readily fill this role. If we then feel contempt for animals for supposedly having immoral desires, we might convince ourselves that we do not harbor those desires.

The error underlying scapegoating is almost always the same whether the victims are humans or animals. Because scapegoating aims to alleviate guilt or shame, when scapegoaters regard the victim, they see the embodiment of repulsiveness or evil. Rather than acknowledge their own weaknesses and flaws, scapegoaters project those very weaknesses and flaws onto victims. Scapegoaters objectify their victims and do not see their victims as individuals who, like them, can feel pain and pleasure and who, like them, desire to live. Instead, scapegoaters see a caricature that greatly magnifies the victims' flaws and fails to recognize both the victims' admirable qualities as well as the victims' experiences as living beings.

Such a view makes it easier for people to maintain a positive self-image while simultaneously endorsing the exploitation and abuse of animals for food, labor, entertainment, and other purposes. Humanity today is probably responsible for more animal suffering and death than at any other point in history.[28] Even though most people say they oppose mistreatment of animals, most tangibly support cruelty to animals. Most people, for example, regularly consume the products of factory farms, where animals typically experience intense, stressful crowding and painful mutilations without painkillers.

People often justify animal exploitation and abuse on the grounds that the animal victims are ugly, dirty, or stupid; that they are merely instinctive; or that they are unable to suffer as people do. However, many of the self-serving distinctions between human nature and animal nature reflect misunderstandings of both.[29] A large body of scientific research contradicts the notion that animals are unthinking and driven entirely by instinct.[30] In fact, many animals display impressive reasoning and communicative skills, form complex social networks, and have rich emotional lives.[31] Finally, there is overwhelming evidence that animals can experience pleasure and pain similar to what humans experience,[32] and even animals who are evolutionarily remote from humans appear to have the capacity to suffer.[33] Indeed, the similarity in feelings, emotions, and many mental attributes between humans and many animals is readily apparent, even to casual observers.

Animal Sacrifices

The language of animal experimenters suggests one way that mistreatment of animals can be quasi-religious. Animal researchers routinely talk of "sacrificing" animals rather than killing them,[34] indicating an element of sacred, scapegoating violence in animal experimentation. In prescientific times, people often made sacrifices to their gods in hopes of restoring their own health or that of someone they loved. These sacrifices could include animal sacrifices, as well as self-sacrifice, such as fasting. Today, in a more secular manner, people often blame nature for illness, and researchers often sacrifice animals to force nature to yield the secrets of health and disease. Ironically, while researchers often sacrifice animals in the name of alleviating human disease, eating animal products contributes substantially to disease in the West.[35]

Animal experimenters play a role analogous to the tribal medicine man, who offers sacrifices to heal the sick. In primal cultures, profane objects—particularly, but not exclusively, animals—become sacred when priests use rituals to transform animals' profane bodies into sacred sacrifices to the divine. By analogy, the sacrifices of animals in modern laboratories purportedly transform the profane (i.e., animal bodies) into what researchers now regard as sacred—data that they maintain will lead to life-saving discoveries.[36] Meanwhile, a scientifically naive public often regards researchers analogously to primal people's view of their medicine men—people whose mysterious activities hold the key to life and death.[37]

There are other contexts in which animals are scapegoating or sacrificial victims. In our pursuit of self-esteem, there are always winners and losers when people compete against each other. However, the ability to dominate or kill an animal can give anyone a sense of self-esteem. A trapping handbook relates, "While many youths develop interest in sports or good grades in school, some do not when they realize that they cannot excel. . . . Any young person, regardless of social advantages, can excel and be an achiever by catching the big fish of the day, or making a nice shot, or catching a mink."[38] There will always be victims as long as the path to self-esteem requires dominating others.

As in all scapegoating, people believe that they are doing something valuable when they sacrifice or scapegoat animals. They blame the victims for being evil: "deer are killing too many people" (from car–deer collisions), "wolves are dangerous, ravenous killers," or "pigs are disgusting." The tendency to exclude from the circle of compassion and concern those animals humans routinely abuse puts all vulnerable individuals, human and animals, at risk. This is particularly the case if, as I argue, the scapegoating process plays an important role in animal abuse and if, as Girard argues, the scapegoating process plays a crucial role in generating human culture and community.

Of course, human motivations are always multifactorial, and scapegoating might, in certain situations, be a relatively small contributing factor to the decision to cause harm. In addition, it is possible for harmful activities to be devoid of a scapegoating or sacrificial element. Some of those who must hunt in order to provide sustenance for their families, a small fraction of hunters in the United States, might genuinely regret the killing but find it necessary.

Must there be losers, human or animal, for people to gain self-esteem? If so, there would seem to be no hope of transcending the tendency to participate in scapegoating violence. Christianity offers an alternative. If God's capacity to love is infinite, we do not need to compete for God's affection. And if God offers love to us unconditionally, we do not need to dominate anyone else to demonstrate our worth. We do not need to prove to ourselves or anyone else that we deserve God's love. If that is the case, our pursuit of self-esteem does not engender scapegoating violence. We can be free from the bonds of acquisitive mimetic desire, rivalry, and destructiveness; and we can become instruments of love and peace. This, I think, was a central message of Jesus.

Many Christians regard the Hebrew Scriptures as important, because they provided the framework for Jesus' ministry. Looking at the Hebrew Scriptures through the lens of Girardian mimetic theory (i.e., mimetic desire leads to mimetic rivalry, which leads to mimetic accusation or scapegoating), these Scriptures offer profound insights into the human tendency to scapegoat and ways to overcome scapegoating's attractions.

Chapter 2: The Hebrew Scriptures

Introduction: My View of the Bible

Because this book focuses considerable attention on the Bible, it is important at the outset to reflect on the Bible's authority. One contentious issue among Christians is whether or not the Bible is inerrant. Some Christians assert that the Bible is a perfect transcription of God's words, and they frequently regard the Bible's inerrancy as a central tenet of Christian faith. They generally justify this stance by claiming that the Bible is remarkably consistent, that it has correctly predicted future events, or that it has reliable witnesses.[1] However, a large body of biblical scholarship contradicts these claims.[2]

Scientific evidence has challenged the biblical inerrancy theory, because a literal reading of the Genesis creation story contradicts voluminous geological, astronomical, and other evidence. The Bible indicates that the earth does not move (1 Chronicles 16:30; Psalm 93:1, 96:10, 104:5), that the sun circles the earth (Joshua 10:13; Ecclesiastes 1:5), and that the earth was created in seven days (Genesis 1:1–2:3). If we were to reject scientific rationality, then logically we would be forced to reject any rational arguments for our religious beliefs. We would then be forced to abandon logic-based and scientific claims for the Bible's truth, such as claims that it is internally consistent, has reliable witnesses, has support from archaeology, or has predicted future events.

Modern scholarship has disputed certain biblical details. Ann Wroe has concluded that nonbiblical texts and archeological discoveries paint a different picture of Pontius Pilate from that in the Gospels.[3] Ancient writers commonly put words in the mouths of real historical people, which evidently was acceptable as long as the writers' accounts pointed toward truths. In ancient times, great spiritual leaders were healers; therefore, if one wished to demonstrate that a person was a great spiritual leader, one generated stories depicting that person healing sick people. Similarly, to gain wider acceptance of their views, authors often used pseudonyms, signing their writings with names of respected writers, such as Paul. Indeed, most scholars now hold that many canonical letters attributed to Paul, including 1 and 2 Timothy, Titus, and possibly 2 Thessalonians and Ephesians, were written by other authors.

Some Christians cite 2 Timothy 3:16 as evidence for the Bible's inerrancy: "All scripture is inspired by God and profitable for teaching, for reproof, for correction, and for training in righteousness." However, when 2 Timothy was written, there was no New Testament; it was canonized in the fourth century. Therefore, 2 Timothy 3:16 could not have been referring to the Bible we use today.[4] Further, only a few of the many gospels and epistles that early Christian churches used and revered were incorporated into the Bible, and during the first 300 years after Jesus' death, great controversies arose as to which documents were authoritative.[5]

There are additional difficulties with the biblical inerrancy theory related to linguistics[6] and to problems inherent in translating from original texts.[7] One does not need to believe that the Bible is inerrant to believe that the Bible's stories, revered for thousands of years, provide profound insights into human nature, community, and humanity's relationship with the earth, animals, and God. Indeed, the Bible's wisdom lends credence to the claim that its authors were divinely inspired. Nevertheless, I acknowledge the possibility that other texts have also received divine inspiration and that the Bible might not be perfectly accurate when it comes to describing historical events.

For me, one of the Bible's most compelling and inspiring overarching messages is a gradual recognition among the Hebrews, and later among Jesus and his disciples, that God is about love and compassion. No doubt Christianity has a violent history, including the Crusades and the Inquisition, but I will argue that such violence has relied on selective readings of the Bible that have been divorced from the Bible's historical and sociological contexts. I will also seek places where the Bible and scientific evidence concur. Though science does not provide absolute certainties, it has proven to be a powerful tool for discerning the workings of nature. As Galileo Galilei (1564–1642) said, "I do not feel obliged to believe that the same God who has endowed us with sense, reason, and intellect has intended us to forgo their use." In my view, the Bible's credibility grows when it accords with scientific findings.

Mimesis in the Garden of Eden

The creation stories of many religions explain the origins of evil and suffering. Remarkably, the Judeo-Christian creation story recognizes the importance of acquisitive mimetic desire in generating conflict and misery. According to this story, Adam, Eve, and all creatures initially lived together peacefully:

> And God said, "Behold, I have given you every plant yielding seed which is upon the face of all the earth, and every tree with seed in its fruit; you shall have them for food. And . . . to everything that has the breath of life, I have given every green plant for food" (Genesis 1:29–30).

There was no violence or death in Eden. God had created Adam and Eve in God's image, and God's desires should have been the model for their desires. God desired harmonious coexistence among all living things, and Adam and Eve should have sought to nurture Eden (Genesis 2:15) and exercise benevolent dominion there (Genesis 1:26–30).

However, the serpent tempted Eve, awakening desires that opposed those of God and threatened the blissful harmony among all of Eden's inhabitants. Adam and Eve, as humans, could not avoid having mimetic desires. However, instead of having God as their model, Eve took the serpent as model, and Adam took Eve. Since Adam's and Eve's models were other creatures, their desires were acquisitive mimetic desires that invariably led to mimetic rivalries. They had become rivals with God for the fruit of the tree of knowledge of good and evil, making it impossible for God to remain their model of love, compassion, and respect. Without respect for God's guidance, they would find it impossible to be loving, compassionate, respectful stewards of Eden.

When God discovered their disobedience, Adam blamed Eve, which constituted scapegoating in that Adam attributed his own guilt to Eve. Likewise, Eve then blamed the serpent. In disharmony from God and each other, God banned Adam and Eve from Eden, which I regard as a metaphorical description of what happens whenever people refuse God's love, crave God's power, and seek superiority over each other. In rivalry with God for dominance in Eden and in conflict with each other for moral justification, Adam and Eve no longer had a loving relationship with God or each other. Concerned that they might eat from the tree of life and live forever (Genesis 3:22), God expelled them from the Garden and forced them to struggle to obtain food, clothing, and shelter. In contrast to the ideal in the Garden of Eden in which Eve was a companion and not an inferior person (Genesis 3:20–25), it was ordained that "he [Adam] shall rule over you" (Genesis 3:16).

In the violent world outside Eden, people have needed social order, maintained by taboos, to reduce divisive mimetic rivalries. However, taboos must be enforced with threats of violence or social disgrace, and therefore taboos have always undermined genuine love and compassion. Similarly, harmonious relationships with animals were broken. The Bible relates that henceforth there would be enmity and violence between humanity and the serpent (Genesis 3:15). This is the tragic, broken world in which we live, a world that can only be fully redeemed by the grace of God.

The Tree of Knowledge of Good and Evil

Why was the tree of the knowledge of good and evil forbidden? This question has prompted much theological reflection. I will offer some thoughts drawn from the social sciences while stressing that I regard the Garden of Eden story as a parable and not as a historical account.

Evolutionary theory presumes continuity between species, and consequently one would not expect there to be any uniquely human attributes. However, certain attributes might be distinctly developed in humans. As humans evolved from prehuman creatures into humans, our remote ancestors came to experience more anger, bitterness, and resentment when they lost competitions for objects of desire. Whereas the desires of prehuman creatures were likely more ephemeral and material, such as immediate satisfaction of hunger or sexual cravings, human desires tend to be more persistent and symbolic, such as gaining self-esteem or a sense of meaning.

As a parable of humanity's origins, I think the Garden of Eden story describes the emergence of what we might call human self-consciousness. While many animals demonstrate self-consciousness,[8] the Garden of Eden story explores the consequences of experiencing human self-consciousness.

Human self-consciousness involves the ability to recognize that one is a living being distinct from others. We can imagine how others perceive us, and we can empathetically imagine how we would feel if we experienced what others experience. This ability to see things from other perspectives allows us to envision different possibilities emanating from a given situation. Therefore, to the degree that prehumans lacked this kind of self-consciousness, they likely were unable to perceive evil, because recognizing evil requires an ability to recognize that other, better possibilities could arise from a given situation. Prehumans surely experienced suffering and fear, but most likely they had little capacity to view these experiences as "evil."

Prehumans tried to avoid pain and death, but they did not seek to understand *why* the world included pain and death, because they could not imagine other possible realities.[9] Without these cognitive skills, prehumans generally experienced the world as resembling the biblical Garden of Eden. They usually had enough to eat, they were not preoccupied with worries about possible future food shortages, and they were not distressed about the prospect of their inevitable demise.

Gaining human self-consciousness, humans saw themselves as actors in a world in which they might suffer or be killed at any time and would definitely die. This describes the effect of eating from the tree of knowledge of good and evil: Knowing good and evil, humans could no longer relate to God and God's creation in harmonious balance. Fearing possible suffering and death, humans came to see the world as full of danger, competition, and strife, even in the absence of immediate dangers or challenges.

Consequently, in times of plentiful food and other resources, people have tended to hoard as a hedge against possible scarcity, often generating actual scarcity. With scarcity, conflicts have arisen that have often led to violence. One thing that has kept people from hoarding excessively has been the fear of being killed by the victims of their rapaciousness. Since fear of death has been one of the few things that has tempered acquisitiveness, eating of the tree of life would have given Adam and Eve immortality and, therefore, power commensurate with God. The only way to save Eden from total destruction was for God to banish Adam and Eve. Therefore, the human desire to distinguish good from evil, a consequence of human self-consciousness and abstract thinking abilities, has made it impossible for humans to experience nature as the mythical Garden of Eden.

It is noteworthy that, after eating the forbidden fruit, "The eyes of both were opened" (Genesis 3:7), which I regard as metaphorically describing their gaining human self-consciousness. With such self-consciousness, Adam and Eve could regard themselves as a third person might regard them, and from this perspective they saw their sexual and other desires as shameful. In contrast, previously they were unabashedly naked and "one flesh" (Genesis 2:24). After eating the forbidden fruit, they recognized their nakedness and covered themselves with sewn fig leafs. In other words, human self-consciousness generates the capacity to feel guilt and shame, and guilt and shame are unpleasant feelings that people are eager to transfer onto other individuals. Consequently, Adam blamed Eve and Eve blamed the serpent for their transgression.

Human self-consciousness allows us to imagine how God views us; and our shortcomings, including our illicit desires, make us feel vulnerable to divine condemnation. Therefore, accompanying human self-consciousness is a strong inclination to transfer our sense of guilt and shame onto other individuals, i.e., to scapegoat.

The First Murder Victim

Girard has maintained that the origin stories of primal religions depict one or more murders that relate to actual killings. Because the murderers are generally the ones who relate the events and because they have believed their actions were justified, the account of the killing has typically asserted the victim's guilt. The killers have seen the murdered individual as an embodiment of evil, not an innocent victim. Indeed, Girard has argued, this has been the purpose of myth—to hide the truth that human culture is grounded on murder, and that murder continues to bind human communities. The Hebrew Scriptures are remarkable in that they frequently recognize the victim's innocence.

The account of Abel's murder illustrates this well. The story begins by noting that God "had regard" for Abel's sacrifice, unlike Cain's (Genesis 4:3–5). Cain's countenance fell, because he experienced mimetic rivalry with Abel. Envious that God preferred Abel's offering, Cain was unable to control his jealous rage. Cain was a "tiller of the soil" (Genesis 4:2) who did not have animals available for sacrifice, and Abel was his only human companion. Consequently, Cain was unable to displace his anger onto an animal scapegoat, which encouraged him to vent his wrath upon Abel.[10] Interestingly, God had said to Cain, "If you do well, will you not be accepted? And if you do not do well, sin is couching at the door; its desire is for you, but you must master it" (Genesis 4:7). In other words, God told Cain that God would judge Cain on his own merit, not on how he compared to Abel. Nevertheless, Cain's acquisitive mimetic desire led to mimetic rivalry that predisposed to sin. He needed to master his acquisitive mimetic desire or suffer tragic consequences. Not heeding God's counsel, Cain killed Abel.

Then, after Cain denied knowledge of Abel's disappearance, God said, "The voice of your brother's blood is crying to me from the ground" (Genesis 4:10). This is the first of many times that the Hebrew Scriptures relate the voice of the victim, which God hears.

Cain, fearing reciprocal (mimetic) violence against himself, said, "Whoever finds me will slay me" (Genesis 4:14). God prevented escalating violence by putting an identifying mark on Cain and declaring, "If anyone slays Cain, vengeance shall be taken on him sevenfold" (Genesis 4:15). People were unable at this point to resist mimetic violence, and only fear of far greater violence could stop the killing. The latter prophets and the New

Testament tended to describe God as preferring love and compassion, rather than threat of revenge, as a path toward peaceful coexistence. However, primitive humanity was not ready for such an ethic.

The Flood

The Bible relates that God delivered the Flood because "the earth was corrupt in God's sight, and the earth was filled with violence" (Genesis 6:11). The earth was filled with violence, because violence is mimetic. If God desired a world of peaceful coexistence, God would need to re-create the world, starting with Noah, his family, and representatives of every kind of animal. After the Flood, God made a covenant with Noah, his family, and all the animals not to deliver another flood (Genesis 9:10, 9:12, 9:15–17). God recognized that there would still be violence, and he told Noah,

> The fear of you and the dread of you shall be upon every beast
> of the earth . . . into your hand they are delivered. Every moving
> thing that lives shall be food for you; and as I gave you the green
> plants, I give you everything. Only you shall not eat flesh with its
> life, that is, its blood. For your lifeblood I will surely require a
> reckoning . . . Whoever sheds the blood of man, by man shall
> his blood be shed" (Genesis 9:2–6).[11]

Noah's taste for flesh came with the curse that the animals, who had once lived harmoniously with humans, now feared and dreaded Noah. God gave Noah permission to eat animals, but many theologians have seen this as a concession, not a command. There are other examples in the Bible in which God permitted certain activities but did not bestow a blessing or endorsement. God allowed the Hebrews to have a king, even though God warned that kings abuse their subjects (1 Samuel 8); and Moses, on behalf of God, permitted men to divorce on account of their "hardness of heart" (Deuteronomy 24:1–4; see Matthew 19:8; Mark 10:4–5). The Apostle Paul wrote that permitted things are not necessarily desirable (1 Corinthians 6:22, 10:23). Although the ancient Hebrews believed that their scriptures permitted meat consumption, they also believed that the blood carried the life force, and the prohibition against consuming blood reminded the Hebrews that all animals' lives ultimately belong to God.

Noah might have been righteous by the standards of his day, but he was far from perfect. After harvesting grapes from his vineyard, Noah got drunk and fell asleep naked. Ham saw his father in this disgraceful state and informed his brothers, and Noah then cursed Canaan, Ham's son. This story illustrates that even Noah, the best of his generation, could be violent, impulsive, and prone to scapegoating in that it would have been dubious to curse Ham and it was unjustified to curse Canaan. Evidently, Noah was prone to violence, and perhaps God allowed Noah to kill animals in an effort to contain his violence. Because God promised to not deliver another flood, God had to give Noah an outlet for his violent tendencies. Does this demonstrate God's indifference to animals? I do not think so. Humanity was uncontrollably violent, and the Bible relates that God hoped that all creation might one day live in peace (Isaiah 11:6-9, 65:25). The first step would be to prohibit killing humans, since animals are far less inclined than humans to hold long-term grudges, seek vengeance, and participate in retaliatory violence.

Abraham and Isaac

The Hebrew patriarchs faced real challenges, sometimes heroically rising to the occasion and sometimes showing poor judgment. Importantly, many stories about them are ambiguous, challenging the reader to discern the story's meaning and whether the participants acted properly. In one such story, God tested Abraham by commanding him to sacrifice his son Isaac (Genesis 22:1–19). Should we admire or condemn Abraham for planning to abide by God's command?

Abraham believed that God expected him to sacrifice Isaac, and many ancient cultures made human sacrifices, particularly child sacrifices, to their gods. If God wanted Abraham to kill Isaac, the God of Abraham would, it appears, have closely resembled the man-made deities to which countless societies have made human sacrifices. Remarkably, there was a radical and dramatic twist; at the last moment, an angel of the Lord commanded Abraham not to kill Isaac. Though there are many ways to interpret this story, I think one reasonable inference is that the story lays a foundation for an understanding of God as one who "desires mercy and not sacrifice" (see Matthew 9:13, 12:7; Hosea 6:6). Abraham then saw a ram caught by his horns in a thicket. Abraham believed that God directed him to sacrifice the ram as a substitute for Isaac, but interestingly the text does not mention that God instructed or wanted Abraham to sacrifice the ram.

Although some regard this story as validating the notion that God some-times desires sacrificial violence, there is ambiguity; Abraham neared the point of sacrificing Isaac, but we cannot know whether Abraham would have carried out the sacrifice if God had not intervened, or whether God would have restrained Abraham's hand if Abraham had begun the motion to cut his son's throat.

Jacob and Esau

The story of Jacob and Esau illustrates acquisitive mimetic desire and mimetic rivalry. It is one of many stories in the Hebrew Scriptures that depict brothers in conflict. As anyone with children knows, siblings almost always fall into rivalry with each other, because their self-esteem depends heavily on how they perform in relation to each other. In this story, Jacob and Esau struggled in Rebekah's womb (Genesis 25:22), and Jacob was born grasping Esau's heel (Genesis 25:26), which I regard as portending Jacob's pursuit of the inheritance and blessing that were rightfully Esau's.

Jacob and Esau's parents spurred the sibling rivalry, with Rebekah favoring Jacob and Isaac preferring Esau. Jacob, with Rebekah's assistance, proved the more cunning of the two, capitalizing on Esau's impulsiveness and shortsightedness to gain both the inheritance and their father's blessing. As is often the case in the Hebrew Scriptures, the younger brother eventually prevailed over the older brother. This undermined notions of the "sacred order," because tradition held that the older brother should have assumed family leadership.

Fearing Esau's wrath, Jacob fled. Later, Jacob's struggle with an angel of God at Jabbok prepared him to meet Esau (Genesis 32:24–31).[12] Once reunited, Jacob bestowed his father's blessing upon Esau and they made peace. As James Williams has observed,[13] this story is about mimetic rivalry resolved without violence. Jacob neither scapegoated nor was scapegoated.

Hierarchies help maintain social order by reducing mimetic rivalries. Consequently, people have often felt that twins threaten the social order. Girard has documented how, in many primal cultures, when identical twins are born, one or both are killed.[14] In our culture, twins are often a source of fascination, if not discomfort. We naturally want to categorize everyone, including children, in terms of looks, intelligence, athleticism, and other features; yet the similarity of identical twins confounds our efforts at differentiation. Lack of differentiation can promote intense fraternal rivalries, and the Jacob and Esau story illustrates the divisive potential of twins.[15]

It is tempting to see the Bible as scapegoating older sons—victimizing them in order to maintain a theme of dominant younger sons. Yet older sons in the Bible often fared well. Ishmael and Esau became patriarchs of great peoples, and Joseph's brothers became patriarchs of the twelve tribes.

Another distinctive feature of the Bible is that the younger sons first endured ordeals in which they were victims (e.g., the near-sacrifice of Isaac, the flight of Jacob from Esau's wrath, and Joseph's travails). Having endured victimization, the younger brothers might have more readily understood the victim's perspective. With this understanding, Jacob's son Joseph showed how love and forgiveness are central to peace and reconciliation.

Joseph

The story of Joseph (Genesis 37–45) revisits the theme of brothers in rivalry, and again the younger brother prevails. Jacob's gift of the multicolored robe to Joseph caused mimetic rivalry and resentment: "When his brothers saw that their father loved him [Joseph] more than all his brothers, they hated him, and could not speak peaceably to him" (Genesis 37:4). Then, Joseph dreamt that his brothers would bow down to him, furthering resentment and anger. Although intense mimetic rivalry can lead to murder, Reuben spared Joseph's life. Reuben failed, however, to prevent his brothers from selling Joseph into slavery. This story illustrates that it is difficult to avoid the pull of mimetic rivalry and mimetic violence, but Reuben showed that love—in his case for his father Jacob, who would have grieved losing Joseph—can sometimes overcome rivalry.

Joseph's adventures in Egypt were also filled with mimetic rivalry issues, but I want to focus on the end of the story, in which Joseph forgave his brothers. The brothers returned to Egypt during the famine, and they did not recognize Joseph, who was in charge of food distribution. Eventually, Joseph forced them to bring the youngest child Benjamin, who was beloved by the father. Joseph framed Benjamin for theft and threatened to enslave the young man. Judah requested that he be enslaved instead, which demonstrated that Judah's love for his father outweighed the envy he likely felt toward Benjamin, Jacob's new favorite. Joseph, evidently moved by Judah's love and contrition and eager to reunite with his family, forgave all his brothers. This forgiveness was critical to his family's reconciliation.

Joseph, who had been treated terribly by his brothers, was able to forgive. He claimed that the events were part of God's plan to "preserve life" and manage the famine. There are similar stories in the New Testament, in which Jesus retained love for those who persecuted and abandoned him, in part because he recognized God's larger plan.

Joshua and Achan

Joshua 7 relates a story about the Hebrews' conquest of the Promised Land. Their repeated triumphs seemed to confirm that God had ordained their land acquisition. Then, they suffered a humiliating defeat at Ai when Joshua, acting on poor scouting information, sent an insufficient number of soldiers to the battle. The story asserts that the defeat was a consequence of someone violating God's command not to take the spoils of victory from Jericho. The story describes God directing the authorities to Achan, and illicit spoils were found in his tent. After Achan confessed to the crime, he and his family were killed and all their property was destroyed.

One can read this story literally and conclude that God ordained Achan's punishment for violating God's command. Or, one can read this story as an example of scapegoating.[16] In support of the latter explanation, as commander-in-chief, Joshua would be held accountable for the debacle at Ai. He was at risk for becoming a scapegoating victim of the sacrificial crisis arising from defeat, unless he could shift the blame. Recall that sacrificial crises lead to a breakdown of the myths, rituals, and taboos. In this case, there was a risk to Joshua that the sacrificial crisis would undermine the taboo against challenging the legitimacy of divinely ordained leaders.

Achan confessed, and gold was found in his tent. However, it is possible that the confession was under duress and that the gold was planted by Joshua's agents. An alternative explanation, which does not involve such cynicism and ruthlessness, was that there were rumors that Achan had taken gold at Jericho. Joshua seized on that knowledge to blame Achan for the military defeat. In this scenario, Joshua might have genuinely believed that Achan was responsible for the disaster at Ai.

Achan was stoned to death, and his family was likewise stoned, perhaps to prevent anyone from coming forward to assert Achan's innocence. Then, Achan's belongings were destroyed, possibly for the same reason that the soldiers were forbidden from plundering Jericho—bitter competition for the spoils could divide Joshua's powerful army.

A literal reading of this story is troubling. It suggests that God is vengeful, not only against the guilty party, but also against his innocent relatives. A Girardian reading posits that Joshua, perhaps maliciously or perhaps not, shifted blame from himself onto Achan. Joshua attributed the accusation to God, which is typical of communal scapegoating, and he utilized mimetic accusation to convict Achan. Once Joshua accused Achan, it is likely that others quickly joined the chorus, eager to find the evil person responsible for the military defeat and possibly also eager to avoid being accused themselves.

This story indicates that the ancient Hebrews, like other ancient peoples, engaged in scapegoating violence. Though the story blames Achan, the author left room for speculation as to whether Joshua was indeed at fault for the military defeat. The beginning of the story relates Joshua's miscalculation, which proved disastrous. Therefore, we can read this story literally as demonstrating that the Hebrews engaged in "divinely ordained" violence, common to all, or nearly all, ancient religions. Alternatively, we may consider that this story exposes "sacred" violence as a scandalous consequence of human scapegoating.

Sacrifice in the Hebrew Scriptures

Although the Hebrew Scriptures often show concern for victims, the prescriptions for sacrifice described in Leviticus present problems for those seeking a nonscapegoating reading of the Bible. Sacrificial violence often involves scapegoating. Likewise, the Hebrew sacrifices also pose difficulties for Jews and Christians who claim that God cares about animals. Why would such a God accept or even encourage killing innocent animals?

A closer look at Leviticus offers some insight. First, Leviticus 1 and 2 repeatedly refer to the proper way to make sacrifices of animals or plant foods *if* one wished to make an offering to God. These chapters do not portray God commanding animal sacrifices. The ancient Hebrews lived among human-sacrificing and animal-sacrificing peoples, and arguably they could not imagine a God who had no desire for blood sacrifices. They were convinced that sacrifice was necessary to approach God in prayer or to appease God after one had transgressed God's laws. Interestingly, the Hebrew law treated all slaughter as sacred sacrifice and mandated the participation of a priest who ensured that the slaughter abided by the humane standard of the day. One who disobeyed this law "shall be cut off from among his people" (Leviticus 17:4).

Leviticus 3:1–16 discusses peace sacrifices. Again, the Hebrew Scriptures describe such sacrifices as optional, elaborating on the proper ritual *if* the sacrifice were an animal. Leviticus Chapter 4 gives instructions for sacrifices if one of God's commandments were unwittingly broken. All primal cultures have feared a sacrificial crisis as a result of violating a divine law, and they have often resorted to human or animal sacrifice in order to appease the divine. Chapter 5 of Leviticus discusses how to make amends for sins against other people and then offer a sacrifice to God. Certain offenses in some primal cultures have called for human sacrifice, and Leviticus 5 may have prescribed animal sacrifices as substitutes. Numbers 31:30 and 31:40 describe human sacrifices ordained by God, and Abraham was on the verge of sacrificing his first-born son to God (Genesis 22:9–13). However, several passages condemn human sacrifice (Deuteronomy 12:31, 18:9–12; 2 Kings 16:3; Psalm 106:38; Jeremiah 19:4–5). Given that child sacrifice was commonly practiced in the ancient world, including among the ancient Hebrews (2 Kings 3:27; Judges 11:30–40; Jeremiah 32:35; Micah 6:7), people would have regarded a prophet who called for an end to all sacrifices as absurd or satanic. I do not see animal sacrifices in the Hebrew Scriptures as God's ideal; they served as a necessary step in the process of ending all sacrifices. Therefore, the Levitican sacrificial code is compatible with a God who has concern for animals and finds all created animals "good" (Genesis 1:21).

As we will see, the later prophets frequently related God's concern for victims and God's opposition to sacrificial violence. However, the Book of Job challenges the notion that God always sides with victims.

Job

The Book of Job offers remarkable insight into the scapegoating process, as well as into the paradox that there is evil and suffering in a world made by God. Satan declared that God's loyal servant Job would curse God if Job lost his family, wealth, and health. God accepted the challenge, and Satan beset Job with a series of calamities. Job's predicament challenges the ancient Hebrew belief that God is both righteous and all-powerful. However, God permitted Satan to inflict misery on Job, which indicates that either God is not righteous or God is not all-powerful.

Job maintained that his treatment had been unjust. Meanwhile, Job's uncharitable friends asserted that Job must have somehow deserved his suffering. They told Job that he must have sinned against God, though Job (and the reader) knew otherwise.

Clearly, Job's "friends" treated him as a scapegoat. The sacred order had been violated; an evidently righteous man had suffered grievous misfortunes. Rather than offer him the solace he needed, his friends cruelly accused him of wrongdoing, despite having no evidence to substantiate their claims. They needed to scapegoat Job to convince themselves that Job, not they, deserved such misery.

Job, convinced that he had been treated wrongfully, demanded an explanation from God. Job was determined to assert his innocence, even if doing so might prompt God to kill him (Job 13:15). God eventually re–sponded to Job but never fully explained why Job had suffered such misfortune. God asserted his power and majesty but did not contradict Job's claims of innocence and unjust treatment.[17] Nevertheless, traditional translations vindicate God by having Job "repent." Jack Miles disputes the Revised Standard Version translation of Job 42:6, in which Job declared, "Therefore I despise myself, and repent in dust and ashes." While nearly all English Bibles have this or a similar translation, Miles describes this translation as "The filament from which hangs the thread from which hangs the entire traditional reading of Job's last words as a recantation . . ."[18] Job maintained his innocence throughout his ordeal, and God failed to meet Job's challenge to either demonstrate Job's sin or admit that God had mistreated him. Therefore, it does not appear that Job needed to repent, and Miles points out that there are reasonable translations of this passage that do not describe Job's repenting.[19]

If we accept the traditional translation, in which Job repented in the face of God's grandeur and mystery, then we avoid the conclusion that God was guilty of wrongly harming an innocent person. However, this approach raises a serious difficulty. We know that God had previously described Job as "a blameless and upright man, who fears God and turns away from evil" (Job 1:8), and we know that Job was afflicted because God accepted Satan's wager. In other words, according to our notions of justice, Job was a victim of great injustice. If God's justice required that Job repent, then it seems that our notions of justice differ markedly from those of God. If that is the case, we will be forced either to reject God's justice or to admit that we have little insight into how we should behave and live. If we choose the first and reject God's justice, we might still abide by God's laws—not because they seem right or just, but because we fear God's wrath. If we choose the second and reject our own sense of justice, there will be little to prevent us from participating in scapegoating.

Bravely, Job rejected neither God nor his own sense of justice: he addressed God respectfully and did not, as his wife counseled (2:9), curse God. Job courageously insisted on his innocence, and God evidently respected Job's claim. Though God never admitted error, God did declare that Job had "spoken of me what is right" (42:7), gave Job "twice as much as he had before" (42:10), and denounced Job's accusatory friends (42:7).

Although the story relates God's attempt to restore Job's family and fortune, Job likely continued to grieve for the children he had lost, and there was no way that God could fully compensate Job for the unjust physical and mental anguish. The Book of Job's resolution indicates that God, as described in the Hebrew Scriptures, is not always just. Indeed, there are stories in the Hebrew Scriptures in which God seems to ordain unjust violence, destructiveness, and death. Did God actually endorse victimization, or did the ancient Hebrews have a limited understanding of God?

Divinely Ordained Violence

Similar to the myths of other religions, the Hebrew Scriptures often describe God as favoring the Hebrews and supporting their wars. Did God actually endorse violence and, if so, what does this say about the God of the Hebrew Scriptures?

The stories of the Hebrews' conquest of the Promised Land seem to contradict the position that God desires love, compassion, peace, and justice. God instructed the Hebrews to drive out the inhabitants of the Promised Land so that the Hebrews would not "learn to follow the abominable practices of those nations" (Deuteronomy 18:9), such as child sacrifice, divination, and sorcery (Deuteronomy 18:9–12). Although God found these practices an "abomination," the *reason* God ordained the destruction of these nations and their people was to prevent the Hebrews from learning from them (see also Deuteronomy 20:18). The subsequent stories of violence are indeed disturbing:

> But in the cities of these peoples that the Lord your God gives
> you for an inheritance, you shall save alive nothing that breathes,
> but you shall utterly destroy them, the Hittites and the Amorites,
> the Canaanites and the Perizzites, the Hivites and the Jebusites,
> as the Lord your God has commanded (Deuteronomy 20:16–17).

These Gentile people might have had erroneous beliefs and practices, but they were sincerely living according to their own faith traditions, and they were not choosing to act sinfully.

The Book of Joshua describes Joshua following these instructions and slaughtering all the inhabitants of Jericho (except Rahab and her family) (6:21), of Ai (8:24–25), Makkedah (10:28), Libnah (10:29–30), Lachish (10:31–32), Gezer (10:33), Eglon (10:34–35), Hebron (10:36–37), Debir (10:38), and other lands (10:40–42). In another series of battles "All the spoil of these cities and the cattle, the people of Israel took for their booty; but every man they smote with the edge of the sword, until they had destroyed them, and they did not leave anything that breathed" (Joshua 11:14). Similarly, in the war against the Midianites, Moses instructed his commanders to kill everyone, including the male children, but to spare "all the young girls who have not known [a] man by lying with him, keep alive for yourselves" (Numbers 31:18).

The violent behavior of the conquering Hebrews was not unusual at that time, but it does violate contemporary views of righteousness and justice. There are many other troubling stories in which God orchestrates what modern readers would consider unjust violence. Elijah instructed the people to kill the prophets of Baal who, though mistaken in believing that their god could rain down fire, were earnest in their faith (1 Kings 18:40–41). Saul lost his favor with God, which eventually led to Saul's death, in part because he failed to carry out the divine order to kill all the people and animals in Amalek (Exodus 17:14; 1 Samuel 15:2–3). Saul had spared the Kenites living there, who "showed kindness to all the people of Israel when they came up out of Egypt" (1 Samuel 15:6).

These troubling stories raise doubts about God's love, compassion, mercy, and justice. There are several possible explanations. One might regard God as tribal, favoring one group of people and having little regard for the rest of creation. However, there are many biblical passages affirming that all creation belongs to God (Psalm 24:1; Isaiah 66:1–2; Colossians 1:16) and that God cares about all creation (Psalm 50:10–11; Job 38–39; Jonah 4:11; Matthew 6:26; Luke 12:6). The Scriptures instruct the Hebrews to show hospitality to strangers (Exodus 22:21, 23:9; Leviticus 19:33–34), and there are many passages that describe God's eventual reign over all the nations of earth (Psalm 22:27; Isaiah 2:2–3, 11:9, 42:6, 61:11; Jeremiah 33:9; Revelation 21:1–4, 21:24). More importantly, such a view of God as favoring a particular tribe makes the Jewish faith, upon which Christianity is grounded, resemble countless other religions that have claimed that their gods have endorsed their violence. There would be little reason to believe that the Hebrew account of the divine is any more valid than myriad other conflicting, self-serving accounts.

Another explanation for God's evident sponsorship of violence against innocent victims is that God is not always just and good. This view could also account for God's role in Job's victimization, but it raises several difficulties: If God is sometimes unjust, how do we know when God is being just? If we do not know, then it would seem that Christian faith is not of much help in guiding our lives.

One possible way to address this problem is to consider that God's desires changed over time. The warrior God who sponsored the brutal conquest of the Promised Land evolved into a loving, compassionate, merciful God, as often depicted by the later prophets and the New Testament Scriptures. This view indicates that the Bible is a story about the growth and development of both God and humanity. God created humanity in God's image, and God gained knowledge, including self-knowledge, by observing humanity's evolution from primitive to more modern forms.[20] God found, evidently to God's surprise, that Adam and Eve would disobey, that humanity would later become so embroiled in violence that God would find no alternative to destroying the earth, that the Hebrews would repeatedly regress to idol worship, and that humanity would seem to find countless ways to frustrate God's desires. As humanity changed, so did God. In the Book of Job, God learned about justice and, presumably, obtained greater knowledge of God's own desires for creation.

An approach that I find attractive is to regard God as unchanging but to view human understanding about God's nature as evolving as humanity has matured. Perhaps the reason that the Bible depicts God endorsing violence is that the Bible, written with human hands, sometimes reflects human desires rather than those of God. In other words, the ancient Hebrews, eager to justify their violence against the inhabitants of land they coveted, attributed their own violence to God. Consequently, the God described by the Hebrew Scriptures is often violent and wrathful; the God of the later prophets tends to be much more concerned about mercy, compassion, and justice; and the God of the New Testament desires peaceful, loving communities.

Violence and the Hebrew Scriptures

Violence is a central theme of the Hebrew Scriptures, and over 1000 passages discuss violence or threats of violence.[21] Rarely, the Hebrew Scriptures describe God as violently destructive for no apparent reason. Uncommonly, there are stories in which God angrily takes out revenge for evildoing. Much more frequently, God hands over evildoers to violent humans, who do the punishing for God. Ezekiel 21:31 describes God's wrath against the Ammonites: "And I will pour out my indignation upon you; I will blow upon you with the fire of my wrath; and I will deliver you into the hands of brutal men, skillful to destroy." One might see this as divine retribution. However, one might reasonably conclude that these accounts, written by the ancient Hebrews, reflected tribalism—a self-serving conviction that God shared their desire for vengeance and endorsed their violence.

It is remarkable that, in about 70 passages of the Hebrew Scriptures, people are punished by the effects of their own sinfulness.[22] The writer of Proverbs observed, "He who digs a pit will fall into it; and a stone will come back upon him who starts it rolling" (26:27), and "A man who is kind benefits himself, but a cruel man hurts himself" (11:17). Similarly, the psalmist wrote, "He makes a pit, digging it out, and falls into the hole which he has made. His mischief returns upon his own head, and on his own pate [head] his violence descends" (7:15–16). These passages indicate that violence ultimately hurts the perpetrator, and the New Testament expresses similar thoughts: "God is not mocked, for whatever a man sows, that he will also reap" (Galatians 6:7).

Nevertheless, the ancient Hebrews often identified themselves as innocent victims, and about 100 of the 150 psalms relate anguish at being "despised" and "hated" by "numerous" and "deceitful" tormentors.[23] The psalmist wrote, "More in number than the hairs of my head are those who hate me without cause" (Psalm 69:4; see also John 15:25); and, in a passage Jesus would later recall (Matthew 27:46; Mark 15:34), the psalmist cried, "My God, My God, why hast thou forsaken me?" (22:1). This psalm then describes the writer as a victim of collective contempt and scapegoating: "But I am a worm, and no man; scorned by men, and despised by the people. All who see me mock at me, they make mouths at me, they wag their heads" (22:6-7).

Remarkably, the Hebrew Scriptures show a growing concern for victims in general, and many prophets identify the tragic plights of orphans, widows, and poor people. (See, for example, Deuteronomy 27:19; 1 Kings 17:20–21; Job 31:16–23; Isaiah 1:17, 1:23; Jeremiah 22:3; Ezekiel 22:4–7; Zechariah 7:10; Malachi 3:5). Perhaps the Hebrews' memories of slavery in Egypt made them more sensitive to the predicament of victims.

Although the ancient Hebrews often understood and articulated the victim's perspective, they retained the universal human desire for vengeance. For example, the psalmist wrote, "O daughter of Babylon, you devastator! Happy shall be he who requites you with what you have done to us! Happy shall be he who takes your little ones and dashes them against the rock!" (Psalm 137:8–9). As with the Psalms, the Exodus account ambiguously seems to portray God as both a sponsor and an opponent of victimization.

Exodus

At first glance, the story of the exodus from Egypt seems to demonstrate God's violence. Many have been troubled by the suffering of the Egyptian citizens and soldiers who were victims of the ten plagues, including the killing of the first-born sons. Why should the Egyptian people suffer so much on account of their hard-hearted Pharaoh? In addition, one could regard Pharaoh himself as a victim, since the text repeatedly attributes his hardened heart to God (Exodus 9:12, 10:1, 10:27, 11:10, 14:8).

The biblical account focuses on how the Hebrews were innocent victims.[24] James G. Williams has noted that this story is unusual in relating in detail their long period of humiliation and harsh oppression as slaves. The injustice was compounded by the fact that they had come to Egypt as a consequence of Joseph, who had saved the Egyptians from famine.

Most ethnic groups have origin stories in which they arise and conquer according to the wishes of their gods. They often worship war gods who favor them and grant them sacred lands. While the Hebrews' story includes the divinely ordained conquest of the Promised Land, their story acknowledges their long period of unjust oppression, leaving open the possibility of envisioning God as caring about victims and desiring justice, mercy, and compassion.

The Exodus story includes a series of substitutions that had the ultimate effect of reducing violence, particularly violence against the innocent. The killing of the first-born was less violent than the previous Egyptian edict to kill all of the Hebrews' male infants. Similarly, the sacrifice of lambs constituted a substitution for human sacrifice. Although substituting animal victims for human victims was still violence, it paved the way toward an ethic of compassion and mercy for all living beings. As long as humans were being victimized, there was little hope of a compassionate ethic toward humans or animals.

A remarkable aspect of the Exodus story is that the Hebrews did not aim to retaliate against the Egyptians, only to leave. Often people have sought revenge as much as their freedom, but the Exodus story has suggested a different approach to injustice.

Interestingly, there are Greek accounts of the Exodus that derive from now-lost Egyptian sources. They related that the Egyptians faced a major crisis because they feared spread of disease from a group of people, evidently the Hebrews. The Egyptians expelled these people, who then established a religious identity. One interesting way in which the Egyptian account differs from that of the Hebrew Scriptures is that the Egyptian story blames the Hebrews for the diseases and then, like the scapegoat sent into the wilderness, the Egyptians send the accused troublemakers away.[25]

The biblical Exodus account contains ambiguities about God's role in violence and destructiveness. Perhaps the ambiguity reflects conflicting views among the ancient Hebrews about God and justice. With the Songs of the Suffering Servant[26] (Isaiah 42:1–7; 49:1–6; 50:4–9; 52:13–53:12), we see a clearer picture of God siding with victims and opposing the scapegoating process.

The Suffering Servant as Scapegoat

The account of the Suffering Servant illustrates the injustice of scapegoating. Isaiah introduced the Servant: "Behold my servant, whom I [God] uphold, my chosen, in whom my soul delights; I have put my Spirit upon him, and he will bring forth justice to the nations" (42:1). The Servant would bring forth justice, but not in the traditional manner of primal religions that had regarded justice in terms of divinely sanctioned retributive violence. The justice brought by the Servant would involve revealing the scandal of scapegoating violence. Isaiah related, "I will give you [the Servant] as a light to the nations, that my salvation may reach to the end of the earth" (49:6).

The text articulates scapegoating clearly. Isaiah 53:2–3 describes how the Servant is ugly and friendless—typical features of a scapegoat. Many cultures regard such people as cursed by the gods. When a sacrificial crisis occurs, the mob often accuses such peripheral members of the community of violating sacred taboos, being possessed by demons, or of casting evil spells. Because scapegoats usually have few friends, people will generally not come to their defense. In the story of the Suffering Servant, the community attributed the Servant's pain, suffering, and death to God. In truth the Servant suffered as a consequence of the people's sinfulness: "Yet we esteemed him stricken, smitten by God, and afflicted. But he was wounded for our transgressions, he was bruised for our iniquities" (Isaiah 53:4–5).

Isaiah reinforced this message: "All we like sheep have gone astray; we have turned every one to his own way; and the Lord has laid on him the iniquity of us all" (Isaiah 53:6). Then, Isaiah reminded his Hebrew people that the Servant was innocent: "he had done no violence, and there was no deceit in his mouth" (Isaiah 53:9). Unequivocally, the Servant was a victim and did not deserve the violence he received.

After acknowledging the Servant's innocence, Isaiah said, "Yet it was the will of the Lord to bruise him; he has put him to grief; when he makes himself an offering for sin" (Isaiah 53:10). It might appear that we should attribute the Servant's misery to God, but Isaiah 42:2 reads, "He will faithfully bring forth justice." The text indicates that the Servant chose to faithfully fulfill his destiny. Consequently, there is a parallel between the Servant making "himself an offering for sin" and Jesus' self-sacrifice. As I will discuss later, it was God's desire, but not God's mandate, that Jesus would allow himself to be a victim of the mob and expose the sin of sacred violence. According to this perspective, God neither desired nor perpetrated the suffering of either the Servant or Jesus. Instead, God desired that the Servant and Jesus would undermine scapegoating violence by showing a path of love and peace that might result in their becoming victims themselves. If God had orchestrated their victimization, God would have become a participant in the scapegoating process. Arguably, the Judeo-Christian God would then resemble the countless other deities who seem to endorse human-inspired, unjust, scapegoating violence.

Why do animal sacrifices not undermine sacrificial violence, when animals are so clearly blameless? Animal sacrifices involve a priest who has the power to transform the mundane—an animal's body—into something sacred that satisfies the divine. An animal sacrifice can represent a "gift" to a god to appease that god for a person's or group's past, present, or future offenses. Or, a priest can perform a ritual that transfers the sins of people onto an animal, and then the priest sacrificially kills the animal. Communities can project their fears, hatreds, or sins onto anyone whom they do not regard as one of themselves, such as animals or people belonging to foreign tribes.

However, during sacrificial crises, there is often the sense that someone within their own tribe has betrayed the group. Almost anyone can be accused, because most people are guilty of some transgression that could serve as a pretext for condemnation. Even for those rare individuals who are above reproach, such as the Suffering Servant or Jesus, the mob can generally find some excuse to justify its violence. It can see a scapegoat's resistance to capture as evil-inspired violence, it can regard a scapegoat's protestation of innocence as a contemptible lie, or it can consider a victim's silence as an acknowledgement of the communal verdict of guilt. As discussed in the next chapter, Jesus did not fall into any of these traps—he did not resist, protest, or remain silent. Similarly, "there was no deceit in [the Suffering Servant's] mouth."

The Isaiah text continues: "The will of the Lord shall prosper in his hand . . . by his knowledge shall the righteous one, my servant, make many to be accounted righteous; and he shall bear their iniquities" (Isaiah 53:10–11). Many commentators have observed that this passage predicts that people will learn from the self-sacrifice of the Servant, and this will lead to righteousness. A Girardian understanding is that the Servant chose to fulfill his divine destiny to reveal the scandal of sacred violence. If the Servant had been guilty in any way, people could have defended his death as divine justice by a wrathful God. However, the Servant's clear innocence shows that he was a victim of scapegoating.[27]

The Later Prophets and Sacrifices

The Hebrew Scriptures describe the paradigm of using animals as scapegoats:

> Aaron shall lay both his hands upon the head of the live goat,
> and confess over him all the iniquities of the people of Israel, and

all their transgressions, all their sins; and he shall put them upon the head of the goat, and send him away into the wilderness by the hand of a man who is in readiness. The goat shall bear all their iniquities upon him to a solitary land; and he shall let the goat go in the wilderness (Leviticus 16:21–22).

There are two components for this prescription for sacrificial atonement for sins. First, the priest confesses the people's sins, and then the priest transfers the sins onto a scapegoat. Micah similarly recognized that atonement requires the acknowledgment of sin. However, he asserted that God does not want sacrifices for sinfulness; instead, God desires righteousness:

With what shall I come before the Lord, and bow myself before God on high? Shall I come before him with burnt offerings, with calves a year old? Will the Lord be pleased with thousands of rams, with ten thousands of rivers of oil? Shall I give my first-born for my transgression, the fruit of my body for the sin of my soul? He has showed you, O man, what is good; and what does the Lord require of you but to do justice, and to love kindness, and to walk humbly with your God? (Micah 6:6–8).

Remarkably, this passage recalls the ancient tradition of human sacrifice, and it also maintains that God does not even want animal sacrifice.
Jeremiah also renounced sacrifices, and he said,

For in the day that I brought them out of the land of Egypt, I did not speak to your fathers or command them concerning burnt offerings and sacrifices. But this command I gave them, 'Obey my voice, and I will be your God, and you shall be my people; and walk in all the way that I command you, that it may be well with you' (Jeremiah 7:22–23).[28]

Likewise, Amos prophesied,

Even though you offer me your burnt offerings and cereal offerings, I will not accept them, and the peace offerings of your fatted beasts I will not look upon. Take away from me the noise of your songs; to the melody of your harps I will not listen. But let justice roll down like waters, and righteousness like an ever-flowing stream (Amos 5:22–24).

Isaiah expressed a similar sentiment:

> What to me is the multitude of your sacrifices? says the Lord;
> I have had enough of burnt offerings of rams and the fat of
> fed beasts; I do not delight in the blood of bulls, or of lambs,
> or of he-goats. When you come to appear before me, who
> requires of you this trampling of my courts? Bring no more
> vain offerings . . . cease to do evil, learn to do good; seek justice,
> correct oppression; defend the fatherless, plead for the widow
> (Isaiah 1:11–13, 16–17).

Proverbs 21:3 relates, "To do righteousness and justice is more acceptable to the Lord than sacrifice," and there are other nonsacrificial passages, including 1 Samuel 15:22, Psalm 51:16–17; Isaiah 66:3, and Jeremiah 6:20. Jesus twice (Matthew 9:13, 12:7) echoed Hosea 6:6, which reads: "For I desire steadfast love and not sacrifice, the knowledge of God, rather than burnt offerings." Throughout his ministry, Jesus, like many of the latter prophets, asserted that God wants compassion and righteousness.

The Ten Commandments

According to Girard, the scapegoating process has been the means by which humans have come together and maintained community during times of crisis. For scapegoating to work, people must not recognize that the victim is far less guilty than they believe. If the lie about the victim's guilt were revealed, scapegoating would lose its ability to maintain peace, order, and communal cohesion. Intuitively, people have always understood that scapegoating is the glue that keeps communities together, and I think this is why so many prophets have been—and continue to be—ostracized or killed. Prophets have often revealed scapegoating, showing that people, in an attempt to transform their own injustice into righteousness, have defended their violence as sacred and divinely ordained.

Insofar as the scapegoating process is concerned, I think we can regard the Judeo-Christian story as we might regard the experience of a growing child, who gradually comes to understand the truth. Prior to the times of the later prophets (circa 800–500 BCE), the ancient Hebrews were unable to fully appreciate the sacrificial process because they were deeply immersed in a world grounded on it. Consequently, they needed rules to help reduce the mimetic rivalries that tend to divide communities and lead to scapegoating violence. The Ten Commandments (Exodus 20:3–17; Deuteronomy 5:7–21) embody these rules effectively.

The First and Second Commandments

The First Commandment (Exodus 20:3; Deuteronomy 5:7), that the Hebrews were to worship only one God, was a radical departure from the polytheism that characterized other ancient religions. For one thing, monotheism made it more difficult for the ancient Hebrews to project their own desires and conflicts onto God. People believing in polytheism could envision their own mimetic rivalries and conflicts as having parallels in the mimetic rivalries and conflicts among the gods. With only one God, it was harder for the ancient Hebrews to defend bitter rivalries or vengeful sentiments by pointing to analogous squabbles among deities.

The ancient Hebrews' monotheistic outlook did not guarantee an end to scapegoating, however, because they still saw God as multifaceted. God could still be angry and jealous, as well as loving and compassionate. Consequently, the ancient Hebrews feared God's anger just as they took comfort in God's general sentiment of love and concern for the Hebrew people.

Despite numerous regressions, the Bible gradually reveals an image of God as loving all creation, from the early Hebrew accounts of God's concern for the "chosen people," to the later prophets who often described God's concern for all victims, to the New Testament stories about Jesus reflecting God's boundless love. Benefiting from the Judeo-Christian revelation, and perhaps aided by the Holy Spirit, today we have opportunities for a broader understanding of God's love than did most people in the past. It is possible that future generations will have an even greater grasp of God's love.

People have always tended to envision their gods in anthropomorphic terms. In other words, people have created gods in their own image, believing that their gods have human attributes and human desires. In contrast, I think that monotheism favors seeing God as having only one essence. Perhaps one reason that the ancient Hebrews were repeatedly drawn to worship pagan gods was that they had difficulty seeing God as having but one essence. Polytheism makes it easier to regard the divine as having diverse and conflicting attributes and desires because each god can manifest a distinctive personality trait. However, I think the common practice of seeing God as a single *person* somewhat misses the point of monotheism, because this view permits people to regard God, like humans, as having many personality traits. Such a god somewhat resembles polytheistic deities, with the varied personalities of polytheistic gods melded into the multiple attributes of one deity.

Girardian theory offers anthropological and psychological reasons for monotheism's importance. Polytheistic traditions can facilitate scapegoating because there are no absolute standards to guide values and behavior. People may pick and choose among a range of deities to admire and worship, and their choices invariably reflect mimesis. One day, people can admire a god known for compassion and mercy, and they may attend to the needs of weak and vulnerable individuals. The next day, agitated by a crisis, they can mimetically follow the crowd into admiring a god known for wrathful vengeance—and proceed to scapegoat those same weak and vulnerable individuals. Monotheism undermines, but does not eliminate, such fickleness.

I believe that the single essence suggested by monotheistic faith is love. Though the Hebrew Scriptures often describe God as wrathful, a recurrent theme is God's love and concern for both the chosen Hebrew people and the rest of creation. (See, for example, Leviticus 19:34; Deuteronomy 7:9; 1 Chronicles 16:34; 2 Chronicles 6:14, 7:3; Ezra 3:11; Psalm 33:5, 100:5; Isaiah 63:7; Jeremiah 9:24; Lamentations 3:22.) The New Testament more clearly depicts God's loving nature. John asserted, "He who does not love does not know God; for God is love" (1 John 4:8). How is God equivalent to love? This question continues to challenge Christian theologians. I think "God is love" means that God is about compassion, caring, and mercy. Christian faith also teaches that God still makes judgments about right and wrong, and we disappoint God when we fall short of our potential. However, I do not believe that a loving God, knowing our frailties, condemns or hates sinners. John wrote, "God is light and in him is no darkness at all" (1 John 1:5). God is light, but there is darkness in the world, in part because we no longer experience life as resembling a Garden of Eden in which all creatures coexist harmoniously. Humans have allowed their own acquisitive mimetic desires to supersede God's loving desire for all creation, and humans have attributed their own violence and scapegoating to the divine.

Although the Bible points to God being about love, Christians have widely disparate images of God. A likely reason is that people tend to see God in ways that provide comfort and reassurance. Consequently, people often think that God's desires align with their own preferences, which helps people believe that they are morally upright and justified in God's eyes. Those eager to wield power, such as dominating husbands, authoritarian parents, or tyrants, often envision God as a ruler who governs sternly and sometimes brutally over his subjects. Alternatively, those who seek to live peacefully and cooperatively tend to regard God as kind, loving, and compassionate. I find this image of God far more appealing and better supported by Scripture. However, I do not think it is possible to reconcile "God is love" with a God who endorses abusing humans or animals. Indeed, only a dark, callous God would countenance cruelty to animals, and it is hard to imagine such a God leading humanity, much less all creation, to reconciliation and peace. Further, I would expect that worshipping such a dark God would harden people's hearts and reduce their ability to resist the temptation to scapegoat humans and animals.

The Second Commandment, prohibiting graven images, discourages the universal human tendency toward idolatry, which involves projecting human attributes onto God. Giving God a humanlike face, as was done by nearly all religions of that time, facilitates seeing God as having human attributes. A humanlike God would tend to share our acquisitive mimetic desires and our thirst for vengeance against those who have offended us. The Bible teaches that we have been created in God's image and likeness (Genesis 1:26), which I understand to mean that we have been imbued with a part of God's essence (Psalm 82:6; John 10:34). Our divine nature gives us the capacity to receive God as our ultimate model for our mimetic desires and behavior. Our desires should not reflect those of fellow humans but rather those of God, and I think Jesus accurately described God's desire: that we love one another (John 13:34, 15:12). When we seek to model our lives on God's love, our faith is truly monotheistic.

The Third, Fourth, and Fifth Commandments

The Third Commandment prohibits taking the Lord's name in vain. One reason for this commandment likely relates to the widespread belief among ancient people that God's name had magical powers. The Hebrews believed that God would and should reward those who were faithful to God: consequently, relying on the use of God's name to gain good fortune was disrespectful of God's judgment and was "in vain."[29]

Another basis for the Third Commandment is that we should respect God, because belittling God is akin to Adam and Eve's sin of falling into rivalry with God. When this happens, God no longer serves as our model. If we disregard God, we tend to ignore rather than follow God's laws, leaving us to establish the law among ourselves. This is problematic, because we are mired in mimetic rivalries, and our laws—unless inspired by an ideal that points to God's love—will tend to exacerbate rather than relieve these rivalries. In other words, laws that are not inspired by love tend to become tools for oppression and abuse: laws designed to address immediate needs, rather than to achieve the highest ideals, tend to express either the acquisitive mimetic desires of those with power or the communal desire to engage in scapegoating. The United States Constitution illustrates how secular laws can readily enshrine injustice. The Preamble describes the goal to "establish justice" and to "promote the general welfare," lofty principles that have been cited over the years to help relieve oppressed people. However, to appease slave-owning states, Article IV Section 2 requires that states return runaway slaves to their owners.

The Fourth Commandment, to keep the Sabbath holy, serves several functions. The Sabbath rest ritually reminds people of God's creativity and goodness. At a more practical level, the Sabbath rest helps to revive the mind and body and, in the end, it probably improves productivity. The Sabbath has historically been a time of prayer, reflection, and study that would normally be overlooked were there no injunction to put aside the many other demands of daily life. This Commandment does not only pertain to humans; animals should also rest on the Sabbath (Exodus 20:10; Deuteronomy 5:14), indicating that, according to the Bible, God's has concern for them.

Girardian theory offers insights into the value of the Fifth Commandment to honor our parents. First, such respect parallels our call to honor God, the ultimate creator. Second, our most intense rivalries often take place within the family, and honoring our parents helps reduce potentially explosive conflicts. As children grow, there is increasing rivalry with parents for power and control. Also, parents generally try to reduce conflict among siblings, and honoring parents encourages children to respect the parents' desire for familial peace.

It seems difficult to reconcile the commandment to honor our parents with James and John leaving their father in the boat to follow Jesus (Mark 1:19–20). Likewise, Jesus instructed James not to bury his father and said, "Follow me, and leave the dead to bury their own dead" (Matthew 8:22; Luke 9:60). This surely shocked many people, because abandoning filial responsibilities was scandalous in the Jewish community. But I do not think Jesus opposed fulfilling such obligations; he was trying to show that service to God is most important.[30]

The Sixth Commandment

The Sixth through Tenth Commandments prohibit those behaviors that are most likely to tear communities apart. The Sixth Commandment is "You shall not kill," and in some translations it is "You shall not murder." Many commentators do not believe that this commandment forbids killing during times of war or killing animals, and many biblical passages describe killing enemies and animals without evident reprobation. However, many have understood this commandment as forbidding all killing, and Jesus evidently concurred. He said, "All who take the sword will perish by the sword" (Matthew 26:52) and "If my kingship were of this world, my servants would fight" (Matthew 18:36). Indeed, the early Christians were pacifists.[31]

What about killing animals? Even though most Jews and Christians today eat meat, wear animal skins, and sponsor killing animals in many ways, many Jews and Christians have refrained from killing animals themselves or by proxy (i.e., having someone else kill animals for them). Interestingly, the first Christians, the Jewish Christians, were vegetarian.[32]

Girardian theory provides strong reasons to apply the Sixth Commandment to animals. As discussed in Chapter 1, animals have always been scapegoating victims. If we seek to abandon scapegoating, it will not suffice to change the victims of scapegoating from humans to animals. As long as people believe that scapegoating violence, which some regard as "justice" or "righteousness," can maintain peace and order, scapegoating animals not only perpetrates injustice to animals, it also puts humans at risk. In times of great crisis, animal sacrifices will seem insufficient to restore order, and humans will become victims of scapegoating.

The Last Four Commandments

The Seventh, Eighth, and Ninth Commandments, which prohibit adultery, stealing, and bearing false witness, all serve to maintain peace. Jesus recognized these activities as sources of conflict and destructiveness. He said, "But what comes out of the mouth proceeds from the heart, and this defiles a man. For out of the heart come evil thoughts, murder [Sixth Commandment], adultery [Seventh Commandment], fornication, theft [Eighth Commandment], false witness, [Ninth Commandment], slander" (Matthew 15:18–19).

The Tenth Commandment forbids coveting, and coveting—acquisitive mimetic desire—fuels divisive mimetic rivalries. The Tenth Commandment does not condemn wanting more; it discourages us from wanting what our neighbor has. Of course, if we did not covet, we would likely be content with far less than we have, and it would be easier to meet the needs of everyone. Humans are mimetic creations, and mimetic desires are universal and unavoidable. The solution is not to eliminate mimetic desire but to find a better model than one's neighbor to admire and attempt to emulate. The

First Commandment instructs us to love God. In other words, our desires should be directed at God, not at our neighbor or what our neighbor has. Therefore, the Tenth Commandment, not to covet, is closely related to the First Commandment, to love God.[33]

We are now ready to consider Jesus' life and teachings. According to Christian faith, these elucidate God's intentions for humanity.

Chapter 3: The Life and Death of Jesus

In the beginning was the Word, and the Word was with God, and Word was God. He was in the beginning with God; all things were made through him, and without him was not anything made that was made. In him was life, and the life was the light of men. The light shines in the darkness, and the darkness has not overcome it (John 1:1–5).

Introduction

As in Genesis, John's creation account does not involve violence. Everything was created as a consequence of a single Word (Greek *logos*), which accords with the monotheistic notion that God has a single essence. How do we come to know this essence?

John's Gospel frequently equates God's revelation with light: "He [John the Baptist] came for testimony, to bear witness to the light, that all might believe through him. . . . The true light that enlightens every man was coming into the world" (John 1:7, 9; see also Luke 2:32; John 1:4–5, 8:12, 9:5; 1 John 1:5). John wrote, "And this is the judgment, that the light has come into the world, and men loved darkness rather than light, because their deeds were evil. For every one who does evil hates the light, and does not come to the light, lest his deeds should be exposed" (John 3:19–20). Jesus would reveal that scapegoating is evil and hidden under the cloak of "sacred" sacrifice.

Jesus' Birth

The scapegoat is typically a peripheral member of the community, and Jesus' humble beginnings followed this pattern. His parents were neither wealthy nor powerful, and he was born in a manger. While his lineage (Matthew 1:2–16; Luke 3:23–38) included King David, there was nothing obvious or distinctive that would foretell his important mission. Indeed,

Jesus told his disciples, "The very stone which the builders rejected has become the head of the corner; this was the Lord's doing" (Matthew 21:42, see also Mark 12:10; Luke 20:17). In other words, people would reject the one who would serve as the foundation of God's plan.

If Jesus had been a child of privilege, he would likely have engendered jealousy, a manifestation of mimetic rivalry. If a mob were to later kill such a person, they might have justified their violence on the grounds that the person was arrogant or did not deserve his privileged status. Or, they could have cited the sins of his wealthy parents, grandparents, or ancestors as excuses to kill him. However, for Jesus to expose the scandal of scapegoating violence, he needed to be clearly innocent, and the Gospel birth stories relate his humble beginnings. The stories describe Jesus revered by the shepherds and the wise men because of his relationship with God, not because of any special position within human society.

John the Baptist

John the Baptist related an essential element in overcoming scapegoating violence. He exhorted, "Repent, for the kingdom of heaven is at hand" (Matthew 3:2). Unless we repent, we will constantly struggle to convince ourselves that our actions are justified, and we will regard our vengeance as "justice."

According to Christian tradition, Jesus was sinless from birth and did not need baptism for forgiveness of sins; he was baptized as an act of obedience to God. John initially balked at baptizing Jesus, but Jesus answered, "It is fitting for us to fulfil all righteousness" (Matthew 3:15). The Bible then relates that Jesus perceived God embracing him:

> And when Jesus was baptized, he went up immediately from the water, and behold, the heavens were opened and he saw the Spirit of God descending like a dove, and alighting on him; and lo, a voice from heaven, saying, "This is my beloved Son, with whom I am well pleased" (Matthew 3:16–17; see also Matthew 17:15; Mark 1:10; Luke 3:22; John 1:32).

When John the Baptist first saw Jesus, he exclaimed, "Behold, the Lamb of God, who takes away the sin of the world!" (John 1:29). What is the "sin of the world?" I think the sin of the world has been scapegoating, which has victimized innocent individuals throughout human history. After Cain's murder of Abel, the world was filled with violence. Even the Flood did not eradicate violence and victimization. Noah, prone to anger and violence, scapegoated Ham, blaming Ham for Noah's own shameful, drunken state. Noah's descendants, sharing his human weaknesses, would invariably participate in scapegoating. Because God had promised not to deliver another flood, God would need a different strategy for taking away the sin of the world.

Many Christians regard the sin of the world as Adam's and Eve's disobedience to God. However, their disobedience was an isolated event, which alone could not constitute the sin of the world, unless it somehow could apply to everyone. A popular Christian theology is that, by an unclear mechanism, humanity has inherited Adam's and Eve's sin, making the sin universal. We will consider difficulties of this theory later in Chapter 12.

Why did John the Baptist proclaim Jesus the "Lamb of God"? Recall that those who engage in scapegoating have generally regarded their violence as sacred and the will of the divine. If Jesus had violently destroyed scapegoaters, then the formerly weak, victimized people would have assumed power. They would have quickly started to scapegoat, because they too would have envisioned their violence as divine justice. The only way to dismantle the scapegoating process, to take away the sin of the world, was to expose it as a falsehood and a scandal. Jesus could not be violent and simultaneously reveal the scapegoating process—to violently oppose scapegoating in the name of God would be tantamount to replacing one form of sacred violence with another. So Jesus had to assume the role of the innocent scapegoat, symbolized by the lamb, himself.[1] By his own choice, he would fulfill God's desire to expose the scapegoating process. This exposure would allow the possibility of reconciliation among victims and victimizers, which was impossible as long as humans, thinking that they were abiding by God's desires, participated in scapegoating.

Normally, people justify scapegoating by pointing to some misdeed by the victim. The victim deserved punishment, they would say, for sinning against God. Given that all of us sin, it is usually easy to find some charge with which to condemn the victim. However, the Gospels describe Jesus as sinless, so the Gospels unequivocally reveal Jesus as a victim.

John the Baptist was uncompromising when it came to truth, and this literally cost him his head when he shamed Herod and Herod's wife. What differentiated John the Baptist's martyrdom from that of Jesus and St. Stephen was that John the Baptist exhibited anger and resentment. Consequently, his death did not fully reveal the scandal of scapegoating, because people could blame his execution on an ill-tempered outburst.

John the Baptist called for repentance of sins, and he baptized with cleansing water. He announced, "He [Jesus] will baptize you with the Holy Spirit and with fire. His winnowing fork is in his hand, to clear his threshing floor, and to gather the wheat into his granary, but the chaff he will burn with unquenchable fire" (Luke 3:16–18). What does this mean? Rev. Paul J. Nuechterlein has suggested that "the fire which should most closely be connected with the Holy Spirit is the Fire of Love . . . a Fire of Love that burns away the chaff of our hardness of heart that keeps us enslaved to the sacrificial fires we continue to project onto God."[2]

Nuechterlein's view accords with John the Baptist's proclamation that Jesus is "the Lamb of God, who takes away the sin of the world," not a judge who would condemn or punish the world. Indeed, Jesus said, "I have come as light into the world, that whoever believes in me may not remain in darkness. If any one hears my sayings and does not keep them, I do not judge him; for I did not come to judge the world but to save the world" (John 12:46–47).

Jesus revealed how to live in service for God. This could involve simple acts of kindness and generosity, but it might mean choosing to be a victim of scapegoating rather than resorting to "righteous" violence. From this perspective, John the Baptist prepared the way for Jesus by cleansing people of sins via repentance and the ritual of baptism. Repentance involves acknowledging one's own sins, which helps reduce the human tendency to judge and condemn other people. The baptism ritual symbolically cleans persons (see Mark 1:4; Luke 3:3), helping them feel worthy to be disciples of Jesus.

Before Jesus could take away the sin of the world, he needed to directly confront and overcome any acquisitive mimetic desires that derived from his human nature. In addition, Jesus needed to show his followers that humans can, aided by a prayerful appeal to the Holy Spirit, transcend their own acquisitive mimetic desires. Therefore, Jesus allowed Satan to tempt him with those desires that most strongly entice humans.

The Three Temptations

Jesus, the man, had human desires. To dedicate himself totally to God, he could not simply repress unwanted desires deep in his psyche, where they might emerge at any time and cripple his mission. Instead, he needed to confront fully and directly the three greatest human temptations: the desire to satisfy one's biological cravings, the acquisitive mimetic desire for power and control, and the desire to feel immortal (Matthew 4:1–11; Luke 4:1–13).

In the desert, Jesus fasted and prayed for 40 days, and many people have found that fasting and prayer help clear the mind to focus on God.

Satan (the nature of whom I will discuss in Chapter 8) tempted Jesus to abandon his focus and address his immediate bodily desire for food. However, Jesus rebuked Satan, quoting Deuteronomy 8:3, "Man shall not live by bread alone, but by every word that proceeds from the mouth of God" (Matthew 4:4; see Luke 4:4).

Satan then appealed to the human desire for self-esteem by offering Jesus all the kingdoms he could see from a high mountain if he would worship Satan. The desire to be a king is an acquisitive mimetic desire, because we frequently crave power and control after regarding other people seeking self-esteem through power and control. Remarkably, referring to all the kingdoms of the world, Satan said, "All these I will give you" (Matthew 4:9; see Luke 4:6). The New Testament does not deny that Satan owns these kingdoms, and indeed the satanic scapegoating process underlies all human kingdoms and other power arrangements. Violence or the threat of violence characterizes kingdoms because an essential component of kingship is the ability of the king to impose his desires on others. Since divine relationships are grounded on love, Jesus' kingdom would be very different from those Satan was offering. Consequently, Jesus rejected this temptation, declaring, "It is written, 'You shall worship the Lord your God and him only shall you serve' " (Matthew 4:10; Luke 4:8).

Third, Satan tempted Jesus to test God by jumping from a pinnacle of the Temple and forcing God to save him (Matthew 4:6–7; Luke 4:9–12). If Jesus were to follow Satan's advice, he would be seeking to conquer his human fear of death. But Jesus already had faith in God's love and goodness; and, by abiding by the biblical prohibition not to tempt God (Deuteronomy 6:16), he rejected Satan's enticement.

The Bible describes Jesus as a leader and a hero, but his heroism is distinctive. Joseph Campbell has described the universal story of the hero, who leaves the community, goes into the wilderness, struggles against dangerous forces, and returns with new, divine knowledge.[3] The story of the three temptations fits this mold, but with an unusual twist: Unlike most such hero stories, Jesus' struggle did not involve violence. Jesus did not physically overcome an external demon or a fierce beast, but rather he conquered those human fears and desires that have always encouraged people to victimize vulnerable individuals. Jesus' ministry would show a way that we, too, can transcend our potentially destructive desires. Indeed, the author of the Letter to the Hebrews, referring to Jesus, wrote, "For we have not a high priest who is unable to sympathize with our weaknesses, but one who in every respect has been tempted as we are, yet without sinning" (4:15). If Jesus could not have been tempted, then he could not serve as a model for us as we struggle with our own temptations.

In Luke, the story concludes, "And when the devil had ended every temptation, he departed from him until an opportune time" (4:13). When would this opportune time be? Jesus was most vulnerable when he was tempted to avoid arrest, prosecution, and crucifixion. Jesus, in rejecting the three temptations in the desert, was now prepared to teach God's message even to the point of death. But being human, he often faced temptation just as we face temptation throughout our lives. Indeed, the Lord's Prayer includes the request to "lead us not into temptation" (Luke 11:4; Matthew 6:13).

James, who some authorities think was Jesus' brother, and other authorities think was Jesus' cousin, understood well the dangers of acquisitive mimetic desire. James wrote:

> But if you have bitter jealousy and selfish ambition in your hearts, do not boast and be false to the truth. This wisdom is not such as comes down from above, but is earthly, unspiritual, and devilish. For where jealousy and selfish ambition exist, there will be disorder and every vile practice. But the wisdom from above is first pure, then peaceable, gentle, open to reason, full of mercy and good fruits, without uncertainty or insincerity. And the harvest of righteousness is sown in peace by those who make peace (James 3:14–18).

In what I consider a beautiful articulation of Girardian mimetic theory, James continued, "What causes wars, and what causes fightings among you? Is it not your passions that are at war in your members? You desire and do not have; so you kill. And you covet and cannot obtain; so you fight and wage war" (James 4:1–2). The solution, as James understood, is to focus on God: "Draw near to God and he will draw near to you" (James 4:8).

Importantly, the three temptations story teaches that not all mimetic desires are bad: Jesus had mimetic desires, but he chose to look to God as the model for them; he did not derive his mimetic desires from humans. It is fortunate that we do not need to reject mimetic desires altogether, because even after our most fundamental biological needs are met, we cannot avoid having such desires. However, when our mimetic desires derive from fellow humans, they become acquisitive mimetic desires that engender destructive rivalries. Instead, we should align our desires with those of Jesus: Christianity teaches that this is possible, because Jesus had a human nature, and he experienced the same temptations that we experience.

Another significant implication of the three temptations story is that Jesus only rejected the temptations; he did not destroy the tempter. As discussed in Chapter 8, Satan is not the cause of sin; the most Satan can do is to awaken satanic desires already within us. If we focus on destroying an external Satan, we will be less inclined to recognize our satanic desires and more inclined to scapegoat others we regard as "possessed" by Satan.

The Passion

After teaching and healing in many communities, Jesus entered Jerusalem on a colt amid an adoring crowd (Matthew 21:7–11; Mark 11:7–11; Luke 19:35–36). Traditionally, a conquering king rode a horse that symbolized power. Instead, Jesus rode on a colt, which showed humility and fulfilled the prophecy: "Lo, your king comes to you; triumphant and victorious is he, humble and riding on an ass, on a colt . . . and he shall command peace to the nations" (Zechariah 9:9–10). One way that Jesus commanded peace was to disrupt the violent Temple sacrifices by turning over the money-changers' tables (Matthew 21:12; Mark 11:15) and liberating the animals (John 2:15). Naturally, this angered the chief priests and the scribes. They sought to kill Jesus, but they refrained from arresting him in public because the crowd admired him (Mark 11:18; Luke 19:48).

In *The Last Week*, Marcus J. Borg and John Dominic Crossan argue that the Temple disruption aimed to undermine the "domination system," in which Temple authorities collaborated with Roman imperialists to ruthlessly exploit the people.[4] Borg and Crossan assert that Jesus did not oppose the Temple in general or the Temple sacrifices in particular. Evidently, they think that an antisacrificial explanation for Jesus' actions would contradict their theory. I think Jesus opposed both the domination system and animal sacrifices as different manifestations of the pervasive scapegoating process.

According to Girard, one way the scapegoating process maintains peace and order is by establishing taboos that prohibit violating the sacred hierarchy. These taboos include who can marry whom, what kinds of work one can do, and what official titles one can have. Jesus recognized the scapegoating process as unjust, because it victimizes vulnerable individuals, such as the infirm or widowed. Similarly, animal victims of sacrifice carry the burden of people's guilt, and several latter prophets denounced this injustice (Isaiah 66:3; Jeremiah 6:20, 7:22; Hosea 6:6; Amos 5:21–22; Micah 6:6–8). Girardian theory posits that the scapegoating process generates the myths, rituals, and taboos that maintain *all* domination systems. If Jesus had opposed only the victimization of people while endorsing the victimization of animals, he would not have undermined the scapegoating process that leaves all vulnerable humans and animals at risk.

People typically regard communal scapegoating as "justice" or "sacrifice" that has divine endorsement. Otherwise, their scapegoating would not generate or maintain communal peace. Remarkably, at the Last Supper, Jesus initiated a new sacrament that did not involve sacrificing any victims. Holy Communion, like sacrifice, is a sacrament designed to bring people closer to each other and to God. However, in using the bread and wine, Holy Communion does not require killing anyone. I think this dramatizes Jesus' rejection of sacrificial, scapegoating violence. However, some have argued that blood sacrifices ended because Jesus, the perfect sacrifice, ended the need for sacrifices to atone for human sinfulness. (But there are difficulties with this theology, which I discuss in Chapter 12.)

What about the doctrine of transubstantiation, which sees the bread and wine transformed into the body and blood of Christ? Transubstantiation posits a mystical transformation and does not involve Christians perpetrating acts of violence against anyone.

Jesus did not avoid his victimization. After the Last Supper, he prayed at the Mount of Olives, "Abba, Father, all things are possible to thee; remove this cup from me; yet not what I will, but what thou wilt" (Mark 14:36; see also Matthew 26:39–44; Luke 22:42). The familiar "Abba," akin to "Daddy," reflects Jesus' intimate relationship with his Father. There was no envy or rivalry between them. Jesus prayed that he would not need to experience suffering and death, but he recognized this as his destiny.

Did God desire Jesus' death? I do not think so. I think God desired for Jesus to show how to build communities based on love rather than on scapegoating violence. Unfortunately, but perhaps inevitably, Jesus' ministry offended many people whose position, power, or sense of order was grounded on the scapegoating process. Jesus' dedication to his destiny put him at great risk of becoming a victim of the scapegoating process. By becoming a victim, he exposed clearly and unequivocally that scapegoating is unjust and scandalous.

The Passion: An Anthropological Look

During the Passion, Jesus assumed the status of the scapegoat victim, which was a pivotal move in revealing the scapegoating process. Jesus told Pontius Pilate, the Roman governor of the province of Judea who presided over Jesus' trial and ordered his crucifixion, "For this I was born, and for this I came into the world, to testify to the truth" (John 18:37). Among the truths revealed by the Passion story is an anthropological understanding of the scapegoating process: Throngs in Jerusalem greeted Jesus with "Hosanna!"

A few days later, throngs shouted, "Crucify him!" The crowd's contradictory behavior calls for an anthropological explanation.

The mob's fickleness illustrates how sentiments are mimetic. When the people hailed Jesus' entry into Jerusalem, their enthusiasm was mimetic. When the people condemned Jesus, their accusatory shouts and jeers were similarly mimetic. These scenes are not difficult to imagine; because television news often gives us images of people caught up in the excitement, whether joyous or angry, of a mob. Perhaps we have recognized times when we have joined the mob, swept away by the group's self-reinforcing emotions.

The means by which the authorities sought to condemn Jesus reveals much about the scapegoating process. Prior to Jesus' arrest, the chief priests and Pharisees deliberated on what to do with him (John 11:47). They acknowledged that Jesus "performs many signs," but they feared that the excitement Jesus had inspired might encourage the Romans to subdue the Jesus movement with force, and this could prove disastrous for the entire Jewish community. Caiaphas advised that they use Jesus as a scapegoat: "But one of them, Caiaphas, who was high priest that year, said to them, 'You know nothing at all; You do not understand that it is better for you to have one man die for the people than have the whole nation destroyed' " (John 11:49–50). This is the logic of sacrifice—that one person should die in order to restore order and peace. Ancient people, not having the benefit of modern psychology and anthropology, did not understand the scapegoating process; and Caiaphas correctly observed that people did not understand how the death of one man could prevent widespread destruction. Scapegoating can be economical in the short term, because it generally requires few victims to prevent widespread outbreaks of violence; but it is costly in the long term, because victims are repeatedly needed. Regardless of the number of victims, scapegoating is always abhorrent, because it is unjust.

Luke's Gospel further reveals the scapegoating process, relating that after the Crucifixion, "Herod and Pilate became friends with each other that very day, for before this they had been at enmity with each other" (23:12). The reason they became friends evidently relates to their common need to maintain order. During the Passover, a holiday that celebrates the Hebrews' liberation from enslavement in Egypt, there was often agitation against Roman rule. Many Jews sought a Messiah who could free them from the yoke of Roman occupation. Many expected that Jesus, who spoke with wisdom and worked wonders, would liberate them; and the authorities, Herod and Pilate, were probably worried that Jesus might lead a revolt. Meanwhile, the chief rabbis were offended by Jesus' disregard for their authority, which could undermine a social order that favored the chief rabbis and could perhaps result in broader social unrest. Interestingly, Pilate did not find fault with Jesus, but Herod was angered when Jesus did not answer Herod's questions. How would Jesus' execution bring two rivals together, particularly when they disagreed on Jesus' guilt? They became friends because their roles in Jesus' execution were mutually beneficial and complementary: Herod declared Jesus' guilt, and Pilate presided over the execution.

Girardian theory helps address certain questions about the Passion that have troubled many Christians. Given the centrality of the Passion to Christianity, what would have happened if the high priests had not requested Jesus' arrest, or if Judas had not betrayed Jesus, or if the Roman authorities had not chosen to condemn Jesus to crucifixion, or if the mob had chosen to release Jesus rather than Barabbas? If the various actors in the story had not played their parts, would Jesus have been spared his tragic destiny, and would the Passion, which contemporary Christians regard as an essential part of Jesus' ministry, have never happened? Perhaps God directed the actions resulting in the Crucifixion, much like a chess player moving the pieces. However, in this case, the Christian story would be like a fictional novel, in which God is the writer, with the only difference being that God makes actual humans suffer, most notably the innocent Jesus.

Do people have free will, in which case it would seem that Jesus might have survived the ordeal in Jerusalem? Or, are people pawns of God's machinations, in which case we might question God's justice and goodness? Girardian theory offers an answer to this paradox: It suggests that the Crucifixion was predictable. Jesus generated a "sacrificial crisis" by several means. He openly violated a wide range of taboos, most notably taboos that held women, poor people, and infirm people in inferior positions. Jesus also challenged the authority of the priests and scribes. Perhaps most provocatively, he threatened the entire sacred order when he undermined the Temple sacrificial cult by turning over the money-changers' tables and liberating the animals. Jesus' defiance of the sacred order sparked public agitation, threatened to foment communal discord, and made his arrest and crucifixion nearly inevitable.

The Passion: Anti-Semitism

A cursory look at history reveals that Jesus' life, death, and resurrection have not stopped Christians from participating in scapegoating. The victims have included people of color, women, homosexuals, people of differing faiths, fellow Christians who have not shared the particular Christian theology of those in power, and animals. Ironically, the Passion, which revealed the scandal of scapegoating, has been an impetus for scapegoating. Many Jews have suffered ostracism or violence because Christians have blamed them for Jesus' death.[5]

Those Christians who have scapegoated Jews have evidently overlooked the fact that Jesus and his first followers, including his disciples, were Jewish. Neither Jesus nor his followers rejected Judaism; instead, they propounded a theology grounded in Judaism. Indeed, it is not surprising that Jesus' ministry found fertile ground among Jews. Judaism had made great progress in the difficult task of revealing the scapegoating process, exemplified best by the Songs of the Suffering Servant and the writings of several later prophets (see Chapter 2).

I turn to two passages that have often been used to justify anti-Semitism. As Jesus carried the cross on his back, he said:

> Daughters of Jerusalem, do not weep for me, but weep for your-
> selves and for your children. For behold, the days are coming
> when they will say, 'Blessed are the barren, and the wombs that
> never bore, and the breasts that never gave suck!' Then they will
> begin to say to the mountains, 'Fall on us'; and to the hills,
> 'Cover us.' For if they do this when the wood is green, what will
> happen when it is dry? (Luke 23:28–31).

Some have interpreted this as a curse, but I think it is a prediction of violence, destructiveness, and misery for those who failed to follow Jesus. Commentators have opined that Jesus was specifically predicting the disastrous Jewish revolt of 66–73 C.E. that resulted in destruction of the Temple in 70 C.E. and the massacre and enslavement of a large number of Jewish people. Indeed, Luke's readers likely related Jesus' declaration to this revolt, because scholars generally agree that Luke was written after the destruction of the Temple.

In the other passage, the high priest admonished the disciples,

> "You have filled Jerusalem with your teaching and you intend
> to bring this man's [Jesus'] blood upon us." But Peter and the
> apostles answered, "We must obey God rather than men. The God
> of our fathers raised Jesus, whom you killed by hanging him on a
> tree. God exalted him at his right hand as Leader and Savior to
> give repentance to Israel and forgiveness of sins" (Acts 5:28–31).

Peter's response initially sounds like blaming: it appears that the disciples had accused the priests of murder. However, Peter then said that Jesus' resurrection was designed to give repentance to Israel and forgiveness of sins to all. In other words, Peter was not trying to shift guilt to the priests, but rather to demonstrate that Jesus' ministry was about repentance and forgiveness.

We, not Jesus or the twelve disciples, are the ones who obsess over blame. Consequently, we tend to scapegoat by attributing far more guilt to the accused than they deserve, thus absolving ourselves of blame. Gil Bailie has observed, "The crucifixion's anthropological significance is lost if responsibility for its violence is shifted from *all* to *some*."[6] In other words, to the degree that Christians attribute the Crucifixion to "the Jews" or to anyone else, the Crucifixion fails to reveal the universal scapegoating process.

The Resurrection: Jesus' Innocence

The Resurrection is a central event in Christianity. One important aspect of the Resurrection is that it unequivocally revealed that Jesus was an innocent victim of manipulative leaders and a deluded mob. In contrast, typical primal myths describe victims as guilty of sowing chaos, casting spells, or violating taboos.[7]

The Bible relates that everyone turned against Jesus. The Roman authorities considered Jesus a troublemaker who threatened the peace, and the Jewish authorities charged that he had blasphemed against the faith by claiming to be the Messiah. The mob, angered that he had failed to liberate them from the Roman yoke, cried, "Crucify him!" Even his disciples abandoned him. Those who participated in Jesus' crucifixion—believing that he deserved an ignominious, painful death—would not expect him to be resurrected and to join God in heaven. The Bible, in relating Jesus' resurrection, sends a clear message that the Roman and Jewish authorities and the mob were wrong about Jesus. He was innocent, and they had participated in his murder.

The Gospels emphasize Jesus' innocence in many ways. Judas, returning the silver reward for betraying Jesus, said, "I have sinned in betraying innocent blood"; then Judas hung himself (Matthew 27:3–5). As the Jewish council sought to turn Jesus over to the Romans, "The chief priests and the whole council sought testimony against Jesus to put him to death; but they found none. For many bore false witness against him, and their witness did not agree. And some stood up and bore false witness against him" (Mark 14:55–57). Pilate (Luke 23:4), the fellow condemned criminal (Luke 23:40–43), and the centurion (Luke 23:47) all asserted Jesus' innocence. Evidently, the mob similarly recognized that an innocent man had been killed: "And all the multitudes who assembled to see the sight, when they saw what had taken place, returned home beating their breasts" (Luke 23:48).

Jesus' encounter with Thomas (John 20:24–28) further demonstrates Jesus' innocence. The Jews believed in the raising of the dead, but this was supposed to happen at the end of time. For Jesus to be raised from the dead in Thomas' lifetime proved Jesus' favor with God, which was only possible if Jesus were innocent. Indeed, Mark's Gospel describes Jesus raised and seated "at the right hand of God" (16:19).

The Resurrection: Jesus' Return to Earth

For many Christians, believing in the Resurrection is a prerequisite for calling oneself Christian. However, many people, particularly in this scientifically oriented age, find it hard to believe that the Resurrection really happened. Meanwhile, a careful comparison of the resurrection stories in the Gospels demonstrates numerous inconsistencies that appear irreconcilable.[8] If the Gospels have inaccuracies about details of the Resurrection, perhaps the Gospels are wrong about the historical validity of the Resurrection itself.

There is no way to determine, scientifically, whether the Resurrection actually happened. However, concern about its scientific proof misses the point about *faith* in the Resurrection. Hebrews 11:1 says, "Now faith is the assurance of things hoped for, the conviction of things not seen." Theologically, it is not necessary to prove that the Resurrection is true in the sense that it physically, materially happened. Two important issues are what it means to *experience* the Resurrection, and whether the various Gospel resurrection stories are *eternally* true: that is, whether they reveal knowledge about God's timeless, eternal nature.

The early Christians experienced the risen Christ as a presence in their lives that transformed their natural human fixation on death to a celebration of life.[9] They no longer feared death at the hands of Roman or other authorities, and they were inspired to courageously spread the good news that Jesus had changed their lives. From this perspective, the Gospel resurrection accounts reflect how these early Christians were spiritually transformed by Jesus' ministry; and they experienced a real relationship with what they perceived as the risen Christ. Whether the Resurrection actually occurred, the disciples and other early Christians *experienced* the risen Christ as a spiritual transformation, and so can we. We can gain a faith in God's love and creativity, which encourages us to believe that God is about life and not death. One implication of this faith is that we will likely find ourselves inclined to regard all God's creation with awe, wonder, and respect, which promotes an attitude of loving kindness in everything we do. Another implication is that faith assures us that, whatever happens to our soul or "self"—our sense of unique identity that we carry throughout our lives—it is not bad. When we perceive death as unpleasant, we tend to experience death's shadow over our lives as punishment for our many sins. This encourages us to try to prove our worth by scapegoating—transferring our sense of guilt to vulnerable individuals. To the degree that we have faith in God's love, we gain faith that God forgives our sins and God cares for soul. Consequently, we are less inclined to try to transfer our guilt—and the punishment that we believe should accompany guilt—to others.

What eternal truths do the Gospel resurrection stories tell us? For one thing, they tell us that God's nature differs from human nature. People often have a strong desire for vengeance, but Jesus did not return to punish those who had wronged him. Jesus did not condemn or abuse his disciples; he greeted them in love and friendship saying, "Peace be with you" (John 20:19). In doing so, he participated in reconciliation, not an endless cycle of mimetic recrimination, accusation, and violence. Therefore, I think one of the eternal truths about the Resurrection is that God is about love and forgiveness, not revenge and hate.

There is another, related eternal truth upon which I have dwelt previously. In the nineteenth century, anthropologists discovered that throughout the world, primal religions were telling remarkably similar origin stories. Typically, these stories described a crisis, a killing, and then peace and reconciliation. Because the end of the "sacrificial crisis" has seemed miraculous, many myths have evolved that describe the resurrection of the murdered victim and his transformation into a God.[10]

Influenced by Enlightenment thought, which tended to view Christianity as mere superstition, many nineteenth-century thinkers concluded that anthropology had confirmed their skepticism about Christianity's stories. Christianity does indeed have the same structure as the primal myths: a crisis, a killing, a resurrection, communal reconciliation, and deification of the murdered victim. However, Christianity's story describes the victim as innocent. The community comes together because people have heard the cock crow, not because they have destroyed the evil in their midst. The eternal truth is that God loves all of creation; God does not join humans in hating those who have been blamed for crises.

Human culture has always tried to reconcile widespread conflicts by scapegoating one or a few individuals. Not only does it exclude some members of God's creation, but scapegoating involves violence and injustice. Godly reconciliation requires people to learn about God's love for everyone. How can this happen? The Bible provides two main approaches that complement each other. One involves explicit instructions, such as the Ten Commandments and the Sermon on the Mount. The other involves demonstration, and the Bible relates how Jesus showed love through his life, death, and resurrection.

Chapter 4: Jesus as Teacher

Blessed Are the Meek

Jesus was a Jew who upheld the Jewish law (Matthew 5:17–20) and the Torah: "But it is easier for heaven and earth to pass away, than for one dot of the law to become void" (Luke 16:17). Jesus' Jewishness means that his teachings related to the Hebrew Scriptures' concern for victims and their gradual recognition of the scapegoating process. Jesus' ministry, including his teachings known as the Beatitudes (Matthew 5:3–11; Luke 6:20–22), further elucidated the scapegoating process. Among the Beatitudes' teachings are insights into how to avoid acquisitive mimetic desire and its consequence, scapegoating violence.

Jesus declared, "Blessed are the meek, for they shall inherit the earth" (Matthew 5:5). How can this happen? Are not meek individuals, human and animal, regularly abused? Jesus assured those who were downtrodden that they would prevail and that their woes would abate. However, it is not clear how this would happen. Some listeners probably envisioned God handing over the reins of power. This accorded with traditional notions of justice, in which people eventually get the satisfaction of vengeance. Indeed, Jesus' claim that the meek will inherit the earth has inspired some Christian liberationists to violently overthrow their oppressors.

Revolutionary violence, however, merely substitutes the perceived righteous violence of one group, that of powerful rulers, with the perceived righteous violence of another group, that of the formerly meek who have obtained power. I do not think Jesus was trying to tell the meek that one day they would have political power. Rather, he was teaching that submission, faithfulness, and love will eventually prevail.

Jesus said, "You are the light of the world" (Matthew 5:14), indicating that discipleship itself is the means by which his followers will prevail. He said, "Let your light so shine before men, that they may see your good works and give glory to your Father who is in heaven" (Matthew 5:16). In other words, in the Beatitudes, inheriting the earth involves a moral and spiritual transformation, not a political or violent one.

Jesus emphasized that followers should be nonviolent in thought and action. He stated, "You have heard that it was said to the men of old, 'You shall not kill; and whoever kills shall be liable to judgment.' But I say to you that every one who is angry with his brother shall be liable to judgment" (Matthew 5:21–22). Jesus continued (Matthew 5:23–24) that one must reconcile with one's brother even before offering a gift at the altar, an indication that making peace is more important than religious observance. Without reconciliation, Jesus explained, conflicts escalate; and such conflicts could result in accusations, court proceedings, and imprisonment (Matthew 5:25).

Blessed Are the Poor

In first century Palestine, people regarded poverty, sickness, or disfigurement as signs of divine judgment. People believed that misfortune reflected punishment for one's own sins or the sins of one's ancestors. It is easy to see scapegoating at work here—mistreating impoverished, infirm, or other marginal members of society, who are the typical scapegoat victims, had a "sacred" flavor in that it complemented punishment by God.

Again, Jesus turned common beliefs upside down. He said, "Blessed are you poor, for yours is the kingdom of God" (Luke 6:20; see Matthew 5:3). As we will explore in Chapter 11, I do not think we should regard the kingdom of God as an otherworldly place where virtuous poor people are rewarded with paradise and malicious rich people get their comeuppance. I see the kingdom of God as a state of connectedness and peace with all God's creation. Such a view would have made sense to Jesus' first century Jewish audience, who would not have denigrated God's earthly creation in favor of an otherworldly paradise. I think Jesus was teaching that those who covet wealth disconnect themselves from the rest of humanity and from God's love. As long as poor people avoid the same, common mistake of coveting wealth and becoming envious and resentful of richer people, they will find it easier than wealthy people to develop genuine and honest interpersonal relationships, to relate to God, and to commune with God's creation. Consequently, they will be blessed with the opportunities to receive and give the blessing of God's love.

For rich people to gain self-esteem through wealth, it is crucial that poorer people envy them. Rich and poor are relative terms, and poor people of one community might have more material wealth than rich people of another. From a mimetic rivalry standpoint, what matters most is how much one has in relation to one's neighbors, not how much material wealth one has. Consequently, rich people tend to flaunt their wealth to generate envy, which validates their success; however, envy can lead to humiliation, resentment, and violence.

Rich people protect their assets with police and military forces that protect their property rights. What do wealthy people do when anger and resentment among poor people grows to the point that poor people threaten to revolt or, in democracies, demand heavy taxation of wealth? Typically, wealthy people try to shift the focus of the anger and resentment onto one or more scapegoats, claiming that the social unrest is due to the activities of evil people who have opposed the sacred order or violated taboos, such as communists or "elite" intellectuals.

To the degree that people victimize other individuals—people or animals—they become less connected to the rest of creation, making them feel more alone in a mysterious, often terrifying universe. I think this is why Jesus said, "Truly, I say to you, it will be hard for a rich man to enter the kingdom of heaven" (Matthew 19:23). Similarly, the writer of Ecclesiastes observed that it is vanity to think that striving for personal gain situates people better in the universe; every living thing shares the same fate of death (Ecclesiastes 1:2–3; 3:19). Jesus said, "Woe to you that are rich, for you have received your consolation" (Luke 6:24).

The parable of Lazarus and the rich man (Luke 16:19–31) is instructive.[1] The rich man had a gate to keep Lazarus and other poor people away. After dying, the rich man suffered in Hades, while Abraham welcomed Lazarus in paradise. Lazarus is a Greek name derived from the Hebrew name *Eleazer* or *Elie'zer* which means whom God helps (Exodus 18:4). We might regard Elie'zer as an emissary from God, and Elie'zer appears in the Hebrew Scriptures as a slave (Genesis 15:2–3), a son of Moses (1 Chronicles 23:15), a prophet (2 Chronicles 20:37), and a martyr (2 Maccabees 6:18–31). Rather than form a relationship with a Larazus/Elie'zer, the rich man had a gate to keep him out. Evidently, the rich man arrogantly believed he had no use for Lazarus, a part of God's creation.

Marked disparity in wealth divides communities and harms everyone. While Jesus was concerned about the plight of poor people (Matthew 19:21, 26:9; Luke 4:18, 14:13), they are blessed to be free of the corrupting effects of wealth. Nevertheless, they are not immune to divisive mimetic rivalries, and they can only receive the blessing Jesus declared in the Beatitudes by opening their hearts to God's love.

Blessed Are You That Weep

Why did Jesus say "Blessed are you that weep now, for you shall laugh" (Luke 6:21)? Perhaps one reason is that those who mourn also experience happiness, because happiness and grief are two sides of the same coin. It is through deprivation and loss that we come to appreciate the blessings we cherish. Further, our remembering that life always involves suffering can help us accept our own suffering with patience and perseverance.

This teaching conveys another truth: Our lives are often restricted and inauthentic, because we spend much of our time and effort trying to avoid suffering. Indeed, our dose of suffering can sometimes increase if we take risks that arise when we live according to our beliefs, values, and goals. Nevertheless, our greatest joys often derive from accomplishments that reflect our deepest convictions. Therefore, if we are willing to risk the possibility of weeping we can also experience great joy.

For example, many of us who mourn animal suffering and death sometimes wish we were not so sensitive to and empathetic with animals. However, the same empathy that causes us to mourn for animals also opens us to the possibility of experiencing joy in our relationships with animals and in knowing that we are helping those who need us. Although our sadness can give way to a sense of despair, Christianity offers a message of hope. We may anticipate the realm of God envisioned by Isaiah (11:6–9), in which all creatures will live peacefully together.

Blessed Are the Peacemakers

Violent people almost always believe that their actions are justified, and violent resistance to their activities only heightens their sense of self-righteousness. However, Jesus said, "All who take the sword will perish by the sword" (Matthew 26:52). The only way to stop the cycle of escalating violence is to be a peacemaker.

Jesus' instructions deviated from traditional teachings. Remarkably, he told his listeners (Matthew 5:38) not to follow the ancient Hebrew "eye for eye, tooth for tooth" rule (Exodus 21:24). The human tendency has been to respond to violence with even greater violence, and the Hebrew rule had likely helped limit mimetic violence. However, this "eye for eye" reciprocal violence was not God's ideal. Instead, Jesus advised, "Do not resist one who is evil. But if any one strikes you on the right cheek, turn to him the other also" (Matthew 5:39; see also Luke 6:29). Along this vein, Jesus said, "Love your enemies and pray for those who persecute you" (Matthew 5:43; see also Luke 6:27). This is the only path toward peace, because retributive violence begets more violence.

Does peacemaking apply to animals? Given that animals belong to God, I am convinced that avoiding cruelty to animals and attending to those in need are forms of peacemaking. Indeed, the Hebrew Scriptures encourage animal welfare (Deuteronomy 22:10, 25:4; Psalm 145:9; Proverbs 12:10), and they mandate that one must rescue an animal on the Sabbath, even if the animal belongs to one's enemy (Exodus 23:4–5). In order for peacemaking to be effective and meaningful, it must be a way of life, not something that one does only when it is convenient. Anytime we close our hearts and minds to the suffering of victims, whether human or not, we become more accustomed to tolerating injustice. I think Dr. Martin Luther King, Jr. was correct when he said, "Injustice anywhere is a threat to justice everywhere."

The Sermon on the Mount

The Beatitudes are part of the Sermon on the Mount (Matthew 5:1–48; see also Luke 6:20–49). I would like to touch on some other components of this sermon that relate to the scapegoating process.

Jesus said that it is better to pluck out an eye or cut off a hand than let that eye or the hand cause a person to sin (Matthew 5:29–30). Although Christians generally agree that we should not take these instructions literally, it dramatizes the serious consequences of desire. Jesus prohibited regarding women with lust (Matthew 5:28), a desire that readily excites mimetic rivalries. Also, regarding women with lust objectifies them into things whose value derives only from their ability to satisfy men's sexual desires. Objectification facilitates victimization, and women have been victims of scapegoating throughout the ages. In many cultures, it appears that men have fear of and contempt for women, who the men believe try to "seduce" them. Many of those cultures encourage men to be strong and unmoved by emotions, and men may feel ashamed when they feel unable to control their lusts. Consequently, the men relieve their own sense of guilt and shame by scapegoating—blaming women for men's desires.

Jesus' sermon then permitted divorce on the grounds of unchastity (Matthew 5:31–32). Otherwise, there would be an adulterous marriage, making a mockery of the sanctity of the institution. From an anthropological point of view, one of marriage's functions has been to reduce mimetic rivalries, because married people should not be candidates for sexual liaisons. Importantly, Jesus' teaching helped protect women from victimization by rejecting the standard of his day, which allowed men to divorce their wives on trivial grounds.

Jesus told his followers to be honest always, not just when they swear (Matthew 5:34–37). Jesus went so far as to discourage making oaths, because to do so suggests that one may be dishonest when not under oath. From a Girardian perspective, reverence for honesty is critically important, in part because scapegoating involves a fundamental lie—that the victim is totally responsible for the social crisis or conflict and that the accusers are totally innocent.

The Great Commandment

When asked which is the greatest commandment in the law (Matthew 22:36), Jesus said,

> You shall love the Lord your God with all your heart, and with all your soul, and with all your mind. This is the great and first commandment. And a second is like it, You shall love your neighbor as yourself. On these two commandments depend all the law and the prophets (Matthew 22:37–40; see also Mark 12:29–31).

Jesus' reply recalls Leviticus 19:18, which reads: "You shall not take vengeance or bear any grudge against the sons of your own people, but you shall love your neighbor as yourself." While the Hebrew passage regards "neighbor" as belonging to one's own people, Jesus had a broader notion of the term. One of Christianity's greatest contributions has been to universalize the Judaic law. When asked who is a neighbor, Jesus gave the example of the Good Samaritan (Luke 10:30–37).

The Hebrews generally despised Samaritans, yet the Good Samaritan was a neighbor to the injured stranger. Today, people generally regard all fellow humans as neighbors worthy of our respect and concern, and many people similarly care about certain animals, such as dogs, cats, and horses. Should we regard all animals as neighbors? To be sure, many animals can be good neighbors *to us*, befriending and protecting us. Should we, likewise, befriend and protect animals? I think so, because we have something fundamental in common: We all have received the spark of life from God. Job said, "In his [God's] hand is the life of every living thing and the breath of all mankind" (Job 12:10). The prohibition against consuming animals' blood (Genesis 9:4) relates to the ancient Hebrews' belief that the blood of all creatures carries the essence of life and belongs to God.

The Bible teaches that God cares about animals, and the psalmist wrote, "Man and beast thou savest, O Lord" (36:6; see also 24:1 and 50:10–11). God saved Nineveh on behalf of its cattle, as well as its people (Jonah 4:11). Indeed, the Bible has numerous passages calling for humane treatment of animals.[2]

Job asserted that all animals reflect God's love and concern:

> But ask the beasts, and they will teach you; the birds of the air, and they will tell you; or the plants of the earth, and they will teach you; and the fish of the sea will declare to you. Who among all these does not know that the hand of the Lord has done this?" (Job 12:7–9).

One reason that God created animals, according to this passage, is that the miracle and diversity of life demonstrates God's creative goodness (see Job 39).

Genesis 1:31 describes God reviewing all creation and declaring it "very good." After the Flood, the Bible relates God's covenant with all creation, including the animals, to not flood the earth again. According to the Bible, in God's eyes humans and animals constitute one community. Mark's Gospel reads, "And he [Jesus] said to them, 'Go into all the world and preach the gospel to the whole creation" (16:15), and the psalmist wrote, "Let everything that breathes praise the Lord! Praise the Lord!" (150:6; see also Revelation 5:13).

Christianity teaches that God cares about all creatures (Matthew 10:29; Luke 12:6), and the science of ecology has shown how living beings depend on each other. When humans forget that God's animals are our neighbors, we tend to abuse our privileges, renege on our responsibilities, scapegoat innocent individuals, and threaten the viability of our own species.

It is important to recognize that Jesus taught that we should love ourselves, as well as our neighbor. This makes sense when we recall that we are creations of God. Some Christian leaders have taught self-loathing, evidently in an effort to combat the human tendency toward narcissism. Other leaders have endorsed narcissism, encouraging people to prioritize satisfying their own desires. This is a welcome message for those with power and money, but it neglects the weak, vulnerable, and poor individuals about whom Jesus also cared. Jesus did not promote either self-loathing or self-aggrandizement. In loving ourselves, we care for our bodies and attend to our needs. In loving our neighbors equally, we do not take self-love to unhealthy extremes.

Loving Our Enemies

Jesus taught that we should to love our enemies (Matthew 5:44; Luke 6:35), though this can be exceptionally difficult. How can we love those who have wounded us badly, or who continue to hurt us or our loved ones? Jesus offered some helpful guidance: he showed us that love was primarily about actions, not feelings. The Good Samaritan story (Luke 10:30–35) does not only describe compassion for the injured traveler; it shows how love involves actively helping those in need. When trying to explain the concept of the "kingdom of God" or the "kingdom of heaven" with parables, Jesus described mutually beneficial relationships or people engaged in constructive, righteous activities (Matthew 13:31–52, 18:23–35, 20:1–16, 25:1–13; Mark 4:26–29; Luke 13:18–21).

Loving our enemies, then, is something to do rather than something to feel. Recognizing that all life comes from God, we may act lovingly toward everything as an expression of our love for God. This is easier to do when we realize that we have all, at various times, been enemies of God—victimizing the innocent to maintain our worldviews, our self-esteem, and our lifestyles.

Humanity has been an enemy of God's animals. The vast majority of those living in the West participate directly or indirectly in massive, institutionalized cruelty to animals. People often say they love animals, but when they sponsor cruelty—for example by eating the products of factory farming—they are not practicing love. Not everyone has affection for animals, particularly animals they regard as dangerous, ugly, or unfriendly. Even if we have trouble finding love for God's animals in our hearts, Christian disciples have a sacred responsibility to God to *act* lovingly toward God's animals. Consequently, harming animals unnecessarily is a rejection of God's love and concern for creation and, therefore, an affront to God.

One difficulty is that we hardly ever regard our violence as violence *per se*, but as "defense" or "justice." Our desire for self-justification is strong, particularly when we crave revenge against those who have hurt us or damaged our self-esteem. The Bible records God saying "Vengeance is mine" (Deuteronomy 32:35), and it is tempting to expedite God's justice by taking vengeance on those whom we believe have wronged us. However, the declaration "Vengeance is mine, I will repay, says the Lord" (Romans 12:19) indicates that vengeance is the proper province of God, not people. Of course, God can repay any way God chooses, which allows us to consider the possibility that God does not want vengeance at all.

Parables

One of Christianity's distinctive features is that Jesus relied heavily on parables. Matthew's Gospel relates, "All this Jesus said to the crowds in parables; indeed he said nothing to them without a parable. This was to fulfill what was spoken by the prophet: 'I will open my mouth in parables, I will utter what has been hidden since the foundation of the world' " (Matthew 13:34–35). Parables lend themselves to diverse interpretations, which is one reason that Christians hold such a wide range of theologies despite sharing a common text. Why did Jesus rely so heavily on parables?

I suggest that reading the Bible through the lens of the scapegoating process yields an explanation. If, as Girard has asserted, the scapegoating process is the foundation of human culture, people would have great difficulty appreciating how the scapegoating process pervades all knowledge, including language itself. If Jesus had spoken directly and had tried to explain how everyone participates in scapegoating, he would likely have been ridiculed or even accused of demonic possession and killed.

Those who talk about the gods demanding sacrifices do not need to talk in parables, because the language of sacred violence is a language people have always understood. Often, the most effective way to communicate that God does not want scapegoating violence is elliptically. Jesus provided a wide range of stories with surprise endings and obtuse sayings. This would encourage people to think in new ways without directly challenging their myths, rituals, and taboos. In this way, they might see "what has been hidden since the foundation of the world" (Matthew 13:35).

This understanding of parables also seems to accord with a passage in Mark: "With many such parables he spoke the word to them, as they were able to hear it; he did not speak to them without a parable, but privately to his own disciples he explained everything" (4:33–34). While Mark describes the disciples as notoriously slow to understand Jesus' ministry, the disciples trusted and believed in Jesus. Consequently, unlike the rest of the community, the disciples were less likely to be scandalized by straightforward explanations of Jesus' teachings. In contrast, regarding his public teaching, Jesus explained, "This is why I speak to them in parables, because seeing they do not see, and hearing they do not hear, nor do they understand" (Matthew 13:13).

Questions Raised by Teaching in Parables

Jesus' speaking in parables raises several challenging questions: First, if it is true that people tend to reject those who, with a prophetic voice, have revealed the scandal of scapegoating, how did antisacrificial teachings by some of the prophets become part of the Hebrew Scriptures? Perhaps the ancient Hebrews were starting to recognize the scapegoating process, in which case these writings resonated with them. The Bible relates that God gave the Hebrews the revelation of monotheism that was enshrined in the First Commandment. If a critically important aspect of monotheism is that it envisions God as having one essence, and if that essence is love (1 John 4:8, 4:16), the ancient Hebrews would have recognized truth in prophets who decried the violence and injustice of scapegoating humans and animals. A God with one essence cannot both love creation and want to see parts of it destroyed.[3]

Second, what did Matthew mean when he said that Jesus spoke in parables "to fulfill what was spoken by the prophet" (Matthew 13:35)? This relates to Psalm 78, in which the prophet Asaph wrote, "I will open my mouth in a parable; I will utter dark sayings from of old, things that we have heard and known, that our fathers have told us" (78:2–3). Asaph then described God's anger at the Hebrews' lack of faith after the Exodus from Egypt, while they lived in the wilderness. The Hebrews experienced much violence and death that Asaph attributed to God. According to Asaph, the people's craving for flesh so angered God that God "slew the strongest of them, and laid low the picked men of Israel" (78:31). This likely relates to Numbers 11:31–33, in which the Hebrews in the desert craved meat, even though there was plenty of manna. God provided abundant quail, and "While the meat was yet between their teeth, before it was consumed, the anger of the Lord was kindled against the people, and the Lord smote the people with a very great plague" (11:33).

Perhaps the term "plague" was meant to describe a rash of violence that occurred when everyone attempted to grab as many quail as possible. It is possible that God was particularly displeased that the Hebrews craved flesh when there was ample, nonanimal manna available. Just as Adam and Eve were not satisfied by the plant foods God had provided in Eden, the Hebrews in the desert desired what God did not want them to have.

According to Asaph, recalling the "dark sayings from of old" will identify the Hebrews' past faithlessness and remind the Hebrews of the importance of keeping God's commandments (Psalm 78:7–8). Such faithlessness had manifested itself as worshipping false gods, such as the golden calf in the desert (Exodus 32:4–8), and not abiding by the commandments not to kill, commit adultery, steal, bear false witness, or envy. These last commandments, as Jesus (Matthew 22:39; Mark 12:31) and Paul (Romans 13:9) recognized, can be summarized as a commandment to "love your neighbor as yourself" (Leviticus 19:18).

Third, how can we recognize our own scapegoating? People have always found it easy to recognize when *other* people scapegoat; it is much more difficult to identify their own scapegoating because they tend to regard their own violence as righteous and just. One way to avoid participating in scapegoating is to listen to the victims. We tend to resist hearing victims' accounts, because doing so can make us aware of our personal failings and our own contributions to the strife that plagues our communities. It is easier to look upon past generations and condemn their victimization (e.g., America's crimes against Native Americans) than to recognize and acknowledge contemporary scapegoating (e.g., America's crimes against animals). Another way to avoid scapegoating is to remain mindful that it happens. If we find that our anger is growing, we must step back, remain as detached as possible, and assess the situation. An excellent strategy is to mentally take the perspective of those with whom we are angry and ask, "How would they describe the situation? How would they defend their actions?"

Fourth, if Christianity has revealed the scapegoating process, why have Christians so often participated in scapegoating, such as against people of color, indigenous peoples, homosexuals, and animals? Christians tend to find scapegoating attractive for the same or similar reasons as non-Christians. However, Christian faith offers ways to resist scapegoating's appeal. For insight, we will look at several parables.

Parable of the Weeds

In the parable of the weeds (Matthew 13:24–30), a servant informs his master that an enemy has sown weeds among his wheat. The master orders them not to pull up the weeds immediately, "lest in gathering the weeds you root up the wheat along with them. Let both grow together until the harvest" (Matthew 13:29–30).

This parable reveals much about the scapegoating process.[4] People have always sought to identify evil and destroy it, and this is how Satan works. Satan convinces us that there is evil in our midst and, in our intense desire to eradicate that evil, we accuse and kill many good individuals along the way. The parable of the weeds instructs us to resist the temptation to try to eradicate anything that might be evil. Doing so allows the good and evil to more clearly manifest themselves over time. Otherwise, the evil we do to ourselves can far outweigh the evil wrought by our perceived enemies.

This describes what often happens when people try to eradicate "pest" animals. The balance seen in nature does not accord with humanity's limitless acquisitive desires. Our material desires can blind us to the harm we cause to God's people, God's animals, and God's earth. In the ongoing quest to meet insatiable human appetites, farmers often try to kill those creatures who reduce farmland productivity or who threaten livestock. Greatly reducing the population of certain pest species often has unpredictable consequences, many of which have proven harmful to humans and to the rest of God's creation.

Parable of the Ten Talents

Many people find the parable of the ten talents (Matthew 25:14–30; Luke 19:11–26) among the most paradoxical. The master castigates the servant who buried the one talent with which he was entrusted, rather than risk losing it in an investment. Jesus explained, "For to every one who has will more be given, and he will have abundance; but from him who has not, even what he has will be taken away" (Matthew 25:29; see also Luke 19:26).

Some people have claimed that the master represents God, and consequently the parable shows the importance of hard work and pursuit of capital gain. However, James Alison argues that the servant's error was not the lack of yield, but rather how he expected his master to treat him.[5] The servant explains, "Master, I knew you to be a hard man, reaping where you did not sow, and gathering where you did not winnow; so I was afraid, and I went and hid your talent in the ground" (Matthew 25:24–25; see also Luke 19:21). Luke makes things more clear, writing that the master says, "I will condemn you out of your own mouth, you wicked servant!" (Luke 19:22). According to Alison's interpretation, the servant's expectation is like the person who believes God is harsh and judgmental. Believing that God wants to punish evildoers, those who readily blame other people for their sins engage in scapegoating, which alienates them from God's love and God's creation. On the other hand, those who have faith in a gracious, loving God will find that their loving actions reap bounteous rewards. They will be more connected to their communities, which will yield material and spiritual benefits. Less inclined to acquisitive mimetic desires, their material "needs" will diminish and they will more easily feel satisfied. Further, by strengthening ties to community members, they will encourage everyone to meet each other's social, material, and spiritual needs. Therefore, those who live by this faith will find greater rewards, and those who reject this faith will lose what little they have.

Many people try to hoard resources in an effort to protect themselves against the vicissitudes of life. In doing so, they increase scarcities and promote mimetic rivalries and conflicts. They are acting like the servant who fears a wrathful God and takes preventive measures that ultimately prove to be self-destructive. Those who have faith in God's abundant love live modestly, share with those in need, and try to ensure that there is enough for everyone. The Bible teaches that such people are already well endowed with faith and will prosper (Matthew 6:19–21, 24–34; Luke 16:10–13).

Parable of the Lost Sheep

A Girardian reading offers some interesting insights into the well-known parable of the lost sheep (Matthew 18:12–14; Luke 15:3–7). In the story, the good shepherd leaves 99 sheep unattended until he finds the one who was lost. A Girardian reading suggests that this parable teaches that we should not sacrifice one individual, even if that individual is foolish, to protect the rest of the community. But if we were to renounce scapegoating, we would risk losing its unifying effects, because scapegoating helps restore peace and order during times of crisis; otherwise, accusations, rivalries, and hostilities tend to escalate and envelop communities in violence. In other words, like the shepherd who risks the flock to save one sheep, Jesus encouraged us to risk communal destruction to avoid killing a single, innocent scapegoat.[6] There can never be reconciliation of God's creation as long as there are victims of human violence.

In Luke's Gospel, Jesus explains the parable of the lost sheep: "Even so, I tell you, there will be more joy in heaven over one sinner who repents than over ninety-nine righteous persons who need no repentance" (15:7). Paul provided teachings that can help us understand this saying. After noting that sin invites God's grace, Paul asks, "Are we to continue to sin that grace may abound?" (Romans 6:1). Paul answers his own question in the next verse: "By no means! How can we who died to sin still live in it?" Just as the 99 obedient sheep were unlikely to wander away when the shepherd sought the missing sheep, those who have truly been reborn in Christ are unlikely to stray from the path of love, compassion, mercy, and righteousness after renouncing the community-bonding benefits of scapegoating.

Parable of the Prodigal Son

From a Girardian perspective, the parable of the prodigal son (Luke 15:11–32) encourages forgiveness, even if doing so sacrifices one's own status and well-being. The story begins with the younger son asking for his inheritance. The father divides his "living" (15:12), which describes the property that is the source of his livelihood, between his two sons. Both sons show little concern or respect for their father, and both regard their father's living as their own possessions. They lack the love and compassion that should bind families, and consequently it is not surprising that later the older brother would find no room in his heart to forgive his younger brother's sins. Instead, the older brother would be self-centered, self-righteous, and judgmental.

The younger, prodigal son takes his inheritance and squanders it. When he returns humiliated and destitute, his father does not condemn him. Rather, the father runs to the son, embraces him, and welcomes him back to the family with a grand party. Conventionally, fathers were expected to walk slowly and erect, and sons were to approach fathers with deference; here the father's behavior conveys love and forgiveness.

The father also lovingly forgives his older son who had just berated the father for welcoming back the prodigal son. The father gently explains the rationale for celebrating the younger son's return and reminds the older son, "All that is mine is yours." By forgiving both sons, the father offers the possibility of familial reconciliation that could not happen if the father were judgmental and punitive.

In ancient Hebrew culture, fathers generally asserted their patriarchal authority and demanded respect for their social position. However, loving personal relationships require mutual respect as individuals, not respect based on social standing. Social standing is grounded in the scapegoating process, while loving personal relationships are unrelated to victimization. Presumably, as young children the sons had once loved their father; the father's showing love for his wayward sons was the only way he could reestablish a loving relationship with them.

"I Have Not Come to Bring Peace, but a Sword"

Even though the parable of the prodigal son features familial reconciliation, Jesus also had this to say:

> Do not think that I have come to bring peace on earth; I have not
> come to bring peace, but a sword. For I have come to set a man
> against his father, and a daughter against her mother, and a
> daughter-in-law against her mother-in-law; and a man's foes will
> be those of his own household. He who loves father or mother
> more than me is not worthy of me" (Matthew 10:34–37).

How would Jesus' ministry divide families? It would not result from people abandoning their Jewish faith in favor of following Jesus, because Jesus said, "Think not that I have come to abolish the law and the prophets; I have come not to abolish them but to fulfil them" (Matthew 5:17). Social discord was an initial, inevitable consequence—but not the goal—of Jesus' ministry. He opposed the hierarchical social order that unjustly marginalized members of society, such as those who were poor, widowed, or infirm. Because the social order helped maintain the peace, Jesus' ministry threatened to disrupt all levels of society, including the most fundamental social unit: the family. However, the Bible teaches that our communities are not peaceful or reconciled as long as they adhere to social customs and laws that rely on scapegoating. It is striking that the parallel passage to Matthew 10:34–37 in Luke includes "Henceforth in one house there will be five divided, three against two and two against three" (12:52). The scapegoating mentality is all against one, with people unified by their hatred of the scapegoat. In contrast, Jesus' ministry threatened to lead to destabilizing conflicts, such as three-against-two conflicts that would be difficult to resolve and could tear families and other social units apart.

To illustrate how exposing victimization can give rise to conflicts, the Christian Vegetarian Association (CVA) has received angry correspondence claiming that it misrepresents scripture. People have charged that the CVA's literature is self-righteous, judgmental, and heretical, even though it explicitly denies that diet determines salvation or that meat eating is inherently sinful.[7]

One likely reason that the CVA engenders such hostility is that it reveals animals as victims of injustice. Not wanting to face this uncomfortable truth, our brothers and sisters in Christ tend to scapegoat animal defenders by ostracism, which divides the body of Christ. However, we at the CVA cannot see how the alternative, to countenance victimizing God's animals in the name of superficial Christian unity, would please God. Because our Christian faith calls us to express love, compassion, and peace, we believe that the CVA should continue its ministry—just as Jesus did not abandon his controversial ministry. We do not truly promote peace if, in the name of getting along with our neighbor, we ignore the suffering that our society inflicts on marginal, often unseen individuals. Such peace is an illusion, just as Jeremiah said of Judah's priests and the false prophets, "They have healed the wound of my people lightly, saying 'Peace, peace,' when there is no peace" (Jeremiah 6:14, 8:11).

We need communities that reject scapegoating as the glue that holds them together. When culturally defined relationships become stumbling blocks to reconciliation, we must be ready to establish new kinds of relationships. Indeed, while Luke 12:52 describes a house divided, Luke 13 features the prodigal son parable, in which the father abandons cultural norms and seeks to reunite his fractured family.

Deviating from social standards violates taboos. This can be dangerous and requires courage and faith. I will discuss Christian faith next.

Chapter 5: Some Thoughts about Jesus

The Knowledge of the Lord

The prophet Isaiah envisioned a time when

> The wolf shall dwell with the lamb, and the leopard shall lie down
> with the kid, and the calf and the lion and the fatling together, and
> a little child shall lead them . . . They shall not hurt or destroy in
> all my holy mountain; for the earth shall be full of the knowledge
> of the Lord as the waters cover the sea (Isaiah 11:6, 9; see also
> 65:25).

What is the knowledge of the Lord?

Girard has argued that central to *human* knowledge is the intuitive
understanding that scapegoating generates communal cohesiveness, and the
myths of many religions show how scapegoating binds people.[1] Humans
have often predicated their violence on the grounds that an angry deity
demands punishment of individuals responsible for the sacrificial crisis, and
future sacrifices become necessary to appease the deity's wrath. Importantly,
Christianity describes a gradual awakening to the view that God does not
want sacrificial violence, nor does God want any harm to come to innocent
individuals. God cares about all of creation, and as the psalmist wrote, "The
Lord is good to all, and his compassion is over all that he has made" (145:9).

I think that if we follow Paul's advice to "be imitators of God, as
beloved children" (Ephesians 5:1), we will be gaining "the knowledge of the
Lord." This knowledge involves people's grounding their mimetic desires
on God's desires rather than on each other's desires. With this knowledge,
we can help reconcile all of creation to the peaceful coexistence that Isaiah
11:6–9 describes. Paul wrote, "The creation waits with eager longing for the
revealing of the sons of God. . . . We know that the whole creation has been
groaning in travail together until now . . . as we wait for adoption as sons"
(Romans 8:19–23). The adopted "sons"[2] will have the knowledge of God,
and they will be instruments of peace. "The creation waits with eager

longing" accords with the phrase in Bishop Thomas Ken's doxology "Praise God, all creatures here below" and with Jesus' charge to his disciples, "Go into all the world and preach the gospel to the whole creation" (Mark 16:15; or "to every creature" [King James Version]). God's ideal involves all creatures glorifying God and living harmoniously with each other.

While Isaiah prophesied hope and anticipation for an eventual harmonious "realm of God," the peace that Jesus offered for those struggling in a harsh, judgmental, and often violent world was different. Jesus said, "Peace I leave with you; my peace I give to you; not as the world gives do I give to you. Let not your hearts be troubled, neither let them be afraid" (John 14:27). Jesus did not offer the peace the world might offer, which would include freedom from strife and danger. I think Jesus offered a sense of peace that could help quell the deepest fears in the human soul—fears related to isolation, damaged self-esteem, and mortality. The process of gaining this sense of peace involves aligning our desires with those of God.

How can we ascertain God's desires, given that many of us find God distant and clouded in mystery? Christians regard Jesus as their model, because God was his model. Jesus taught that it was possible to love everyone and avoid victimizing anyone. Many Christians believe that Jesus took the role of the "little child" described in Isaiah 11:6, who will lead all of creation to peace and harmony.

Does one need to be versed in Girardian mimetic theory in order to comprehend the knowledge of the Lord? No: Jesus said, "Follow me" (Matthew 4:19, 8:22, 9:9, 10:38, 16:24, 19:21, and throughout the Gospels). He advised his disciples to simply imitate him, which is easy for mimetic creatures such as ourselves. He assured his followers, "My yoke is easy, and my burden is light" (Matthew 11:30); but Christians today still struggle to understand how this is so. Christians who genuinely try to follow Jesus often suffer from misfortune or from the hard-heartedness of other people, including other Christians. It takes great faith to believe that Jesus' path of love and peace will result in a sense of inner peace and well-being. Lack of such faith, I think, largely explains why Christians throughout the ages have so often engaged in violence and destructiveness.

Is Jesus the Only Way?

The Bible describes many non-Jews and non-Christians as righteous, including Job, Ruth, the Good Samaritan, and the centurion who declared Jesus innocent (Luke 23:47). Nevertheless, John 14:6 reads, "I am the way, and the truth, and the life; no one comes to the Father, but by me." Largely on account of this passage, many Christians believe that only Christians can "come to the Father." I am skeptical of the claim that only Christianity shows the way to the Father. If this were the case, it seems to me that children of Christian parents would have an unfair advantage. Are those who have little or no exposure to Jesus condemned? I think that putting John 14:6 into context offers insight.

Jesus was talking to Jews, many of whom were so focused on the *letter* of the law that they had forgotten the *spirit* of the law. He aimed to reform Judaism to its original intent, which included serving God by reflecting God's love for all of creation. For this audience, following Jesus really was the only way to fully understand God. Importantly, during Jesus' lifetime, Christianity did not exist as a separate religion, and Jesus did not seek to abolish the Jewish law (Matthew 5:17).

All Christians eager to spread the gospel should be mindful that, when they show love and compassion, their activities encourage non-Christians to consider the faith. However, it in no way lessens the validity or importance of Christian teachings to acknowledge the possibility that non-Jews and non-Christians in other places and times have received divine revelations and have known "the knowledge of the Lord" (Isaiah 11:9, 65:25). Indeed, Jesus' stories and metaphors likely have been incomprehensible to some people with different cultural traditions. This is why Christian communities emanating from missionary work often exhibit a melding of Christian traditions with local stories, beliefs, and customs; because Christianity can only make sense to them if put in the context of their own culture.

Christianity's survival and growth have been historical phenomena. The stories about Jesus could have died out, joining countless religious movements that have been forgotten by history. The disciples could have failed to convince people to follow Jesus, or Christians could have abandoned the faith rather than face persecution at the hands of the Romans. Given the wide range of Christian beliefs and practices in the first two centuries after Jesus' death, it is also possible that a Christianity very different from what we have today could have prevailed. If this had happened, Christianity's holy text would likely have included many books and letters that were once condemned and destroyed as heretical; and it would have omitted writings that Christians cherish, or struggle with, today. If the movement inspired by Jesus had died out, or if it had been altered so much as to be practically unrecognizable compared to Jesus' actual ministry, would the world have been deprived of Jesus' teachings? I do not think so, because "God so loved the world that he gave his only Son" (John 3:16). It seems reasonable to believe that a loving God would give the world the Son as many times as needed to reveal divine love.

Some hold that God engineered the success of the Christian movement, but this theory is problematic. If God has had such control, it would have deprived humanity of free will. Faith, obedience, and service would then be acts of God more than acts of human agency. If people's devotion to God were actually directed by God, it would be meaningless. Also, if the way that God has worked has been to manipulate history, then why has God allowed such suffering in the world? If God could make people follow Jesus in the face of Roman persecution, why instead did God not make the Romans abandon their persecution of the Christians? If God directed the success of those early Christians who practiced what we call "orthodox" Christianity, why did God permit "heretical" early Christian sects to flourish in the first place? More generally, why has God allowed people to kill each other, torture animals, and degrade the environment if God has always had the desire and power to stop human destructiveness?

I do not think God controls human decision making. Consequently, Jesus needed to use a range of strategies to convince people to follow his teachings, including direct instructions in the Sermon on the Mount and elliptical teachings in the parables. Toward those who opposed him, Jesus used harsh words:

> Jesus said to them, "If God were your Father, you would love me, for I proceeded and came forth from God; I came not of my own accord, but he sent me. Why do you not understand what I say? It is because you cannot bear to hear my word. You are of your father the devil, and your will is to do your father's desires. He was a murderer from the beginning, and has nothing to do with the truth, because there is no truth in him" (John 8:42–44).

Jesus denounced those who used the law to justify their misdeeds and violence. As long as they resisted Jesus' revelations, they would continue to participate in scapegoating.

The Divinity of Jesus

Christian tradition has taught that Jesus had both a human and divine nature. Jesus said, "I and the Father are one" (John 10:30) and similarly, "The Father is in me and I am in the Father" (John 10:38). Because the Father and Son were in each other, there was no rivalry between them. Therefore, Jesus' loving, nonrivalrous relationship with God has been a model for human relationships with each other. Girard has termed Jesus' striving to mimic the Father as "good mimesis," and Christians are similarly called to good mimesis, with Jesus as our model.

If Jesus were only divine, we would be caught in a bind: We would be taught to imitate Jesus, but he would exist on such an exalted plane that we would have trouble identifying with him as a model. Jesus would not have experienced our challenges and struggles, and therefore his prescriptions for living would have seemed unreasonable. However, Jesus was genuinely tempted by Satan (Matthew 4:1–11; Luke 4:1–13) with the same desires that we often have. Therefore, the writer of Hebrews said, "For because he himself has suffered and been tempted, he is able to help those who are tempted" (Hebrews 2:18).

Jesus was endowed with the full complement of human attributes, including human desires. The important implication is that we, like Jesus, have the freedom to renounce our own acquisitive mimetic desires and to dedicate ourselves to God. Like Jesus, we can have a divine nature; and Jesus said, "Is it not written in your law [Psalm 82:6], 'I said, you are gods'?" (John 10:34). The Holy Spirit that descended upon Jesus at his baptism "like a dove" (Matthew 3:16; Mark 1:10; Luke 3:22; John 1:32) can similarly inspire and guide our lives.

This theology does not diminish Jesus. The mind of Jesus that Christians are called to imitate (1 Corinthians 2:16; Philippians 2:5) is the mind of God. If we believe that we are capable of being partly divine, this does not reduce Jesus' divinity. Indeed, concern about who is most divine is only an issue in a world defined by mimetic rivalry.

Many people claim that their deepest religious convictions derive from divine inspiration. However, the human capacity for self-deception is strong, and it is tempting to believe that our acquisitive mimetic desires reflect the will of God. One way our acquisitive mimetic desires differ from divine desires, I am convinced, is that divine desires are always loving and compassionate. We can be confident that we are on the right path when *others* regard our actions as compassionate. Conversely, when people claim that we are victimizing other individuals, we should seriously consider their concerns.

Miracles

Some Christians have doubted the Bible's miracle stories, in part because they diminish the importance of faith. These skeptics point out that, if Jesus had really performed miracles, following him would be the only rational choice and not a matter of faith. It would be prudent to follow Jesus not because his teachings were good or true, but because he had unequivocally demonstrated his authority.

Another reason some people have questioned the historical validity of the miracle stories is that the accounts seem too incredible for contemporary, scientifically oriented minds. Whether or not the miracles actually happened, the miracle stories repeatedly point to central lessons of Jesus' ministry. John's Gospel consistently refers to Jesus' miracles as "signs," because they had instructional value. As discussed in Chapter 9, many of the miracles involved healing, and they highlighted the importance of the holistic healing of the body, mind, and soul. Complete healing restores a sense of wholeness, which requires more than fixing malfunctioning body parts. Healing involves restoring a sense of connection to the community and to the source of life and meaning, which Christians identify as God. Jesus demonstrated such healing by curing the woman with perpetual bleeding and then declaring, "Your faith has made you well" (Mark 5:34). Other miracles, such as walking on water, illustrated the power of faith.

Jesus' miraculous feeding of 5,000 by multiplying the fishes and loaves of bread is another story that points to an important teaching. John's Gospel relates that the people were becoming hungry, but it is hard to imagine all these people leaving home and traveling long distances without food. More likely, many did not plan well and were hungry. Then a "lad" shared his food, and subsequently there was enough for everyone (John 6:9-11). Evidently, Jesus inspired this lad to share, which was quite miraculous given that scarcity tends to encourage people to hoard. Subsequently, everyone was fed, perhaps because the lad's example encouraged others to share.

Although our culture has generally regarded the world in terms of scarcity, this teaching offers a different view. In our culture, we see competition in sports, the marketplace, and life in general as a zero-sum game: For every winner, there is a loser. We are taught that we should strive to be a winner, but the unfortunate consequence, according to this thinking, is that our gain is someone else's loss. Thinking otherwise is considered "unrealistic" or "silly," and "Nice guys finish last" is a popular aphorism. A competitive attitude encourages us to accumulate wealth and to adopt a win-at-all-costs attitude in pursuit of money, power, and resources.

Jesus showed a different way. He asserted that God's love and grace are abundant, so we do not need to worry about scarcity. If we show love for each other, there will be enough for everyone. Whether Jesus actually miraculously multiplied the fishes and bread is not critical to the story. What is undeniable is the miraculous change in the people; they came to realize that compassion and sharing result in enough for everyone.[3]

Notions of scarcity have often been used to defend both human and animal mistreatment. Wildlife managers often justify "predator-control" programs by blaming predators for falling populations of certain wild animals. In truth, wild animal populations have existed in balance for eons, and human activities such as hunting and fishing have often been responsible for recent population disruptions. Similarly, humans have caused considerable harm to the environment upon which all of us depend. Although people often justify these activities as "necessary," pursuit of more luxurious lifestyles has accounted for much environmental damage.

Jesus taught that compassionate, caring communities find ways to meet everyone's needs. But it requires faith to believe that we can work together as communities to find peace and well-being.

Chapter 6: The Pauline Letters and Christian Faith

"The Wrath of God"

Many people struggle to reconcile the temperament of God as described in the Hebrew Scriptures with that of God in the New Testament. Many passages in the Hebrew Scriptures portray God as angry or even wrathful, but many other passages, particularly in the writings of many of the latter prophets, also depict God as loving, compassionate, and concerned about victims. Micah asked, "What does the Lord require of you but to do justice, and to love kindness, and walk humbly with your God?" (6:8). The New Testament shows Jesus as loving, compassionate, and forgiving. The only time Jesus appears to have acted in anger was when he turned over the money-changers' tables, drove them out of the Temple, and liberated the animals (John 2:15). Even then, Jesus did not injure anybody.

Is God multifaceted, sometimes inclined toward anger and wrath, and at other times inclined toward love and compassion? Many Christians think so, but I regard this understanding of God as problematic, in part because it contradicts the view that monotheism sees God as having a single essence. Rev. Paul J. Nuechterlein has argued that a principal reason Christians regard God as wrathful results from a misunderstanding of "the wrath" in Paul's letter to the Romans.[1] In this letter, nearly all English Bibles have translated the Greek wording for "the wrath" as "the wrath of God" or "God's wrath." But the Greek text does not use the word "God," so attributing the wrath to God reflects translators' assumptions about what Paul meant.

Why is this important? For centuries, Christians, seeing God as wrathful and vengeful, have been tempted to assist in "God's work" by meting out violent "justice" against perceived wrongdoers. In theory, God is fully capable of meting out whatever vengeance God might desire (Deuteronomy 32:35; see also Hebrews 10:30). However, when people believe they have been wronged, they frequently conclude that wrongdoers have violated God's laws. Eager to see "God's vengeance" satisfied, people have been inclined toward "righteous" violence. But is God really vengeful, or does God have only one essence, which is love (1 John 4:8, 4:16)?

Paul used the word wrath (*orgé*) ten times in Romans. The first time (Romans 1:18), Paul actually wrote "wrath of God"—but not once after that. In Paul's time, Jews and early Christians generally attributed calamities and general strife to God's wrath, so it was reasonable for Paul to introduce *orgé* in association with God. However, Paul then quickly clarified his position by showing that the human suffering associated with wrath was actually a consequence of human action. In Romans 1:24, 1:26, and 1:28, Paul described how God "gave up" people to the consequences of their idolatry of worshipping human desires rather than God.[2] In other words, in Romans 1:18, Paul introduced the well-known topic of "the wrath of God," because his readers believed that human suffering was a consequence of God's anger. However, Paul next argued that human misery was actually a consequence of human activities. After Romans 1:18, Paul repeatedly described conflict and misery as "the wrath," and he did not attribute the conflict and misery to God.

I will look at the first seven verses of Romans 3 closely, because they reveal much about Paul's theology about the wrath. Paul wrote, "Then what advantage has the Jew? Or what is the value of circumcision? Much in every way. To begin with, the Jews are entrusted with the oracles of God" (Romans 3:1–2). Here, Paul reminded readers that the law was what God used to exert divine will prior to the arrival of Jesus. Jews, having been entrusted with the law, had a special mission and privilege.

Paul continued (Romans 3–4),

> What if some were unfaithful? Does their faithlessness nullify the faithfulness of God? By no means! Let God be true though every man be false, as it is written, "That thou mayest be justified in thy words, and prevail when thou art judged."

Paul had begun his letter to the Romans by discussing how everyone sins and fails to fully live according to the law. Yet even when Jews were unfaithful, God remained faithful to the Jews: God's promise to them was honored in spite of their faithlessness.

Now we come to a key verse, Romans 3:5: "But if our wickedness serves to show the justice of God, what shall we say? That God is unjust to inflict wrath on us? (I speak in a human way.)" Paul said that, because we are inevitably wicked, God's justice prevents God from condemning us. Yet if this is so, is God unjust to inflict wrath on us? Paul then notes that this is a *human* way of speaking about it. This "human way" of thinking attributes the wrath to *God*, which is what people have been doing since the beginning of human culture. People have always attributed their misfortunes to God or the gods, and they have offered sacrifices to appease the divine.

Romans 3:6–7 reads, "By no means! For then how could God judge the world? But if through my falsehood God's truthfulness abounds to his glory, why am I still being condemned as a sinner?" Because we are all sinners, God cannot judge the world according to our sinfulness. Despite our sinfulness, God's truth abounds in God's glory. What is God's truth and glory? I think God's glory involves creative goodness, and the truth is that God does not want suffering or violence in God's creation (see Romans 8:18–22).

God's love of Paul, a sinner, exemplifies God's love for all of creation, because only God's loving forgiveness—not Paul's actions—can justify Paul. Why did Paul say that he was condemned as a sinner? It was because he lived among humans who were judgmental and vengeful. The wrath is related not to God's violent hand but to God handing humans over to the consequences of their idolatries and wickedness (Romans 1:24–28). Humans, failing to reflect God's love and forgiveness, condemned Paul. Humans, not God, have always been eager to punish anyone they have regarded as a sinner.

Another passage dealing with wrath is Romans 9:22 which reads, "What if God, desiring to show his wrath and to make known his power, has endured with much patience the vessels of wrath made for destruction . . ." In other words, Nuechterlein has argued, "the vessels of wrath made for destruction" reflects human destructiveness and includes things like the whip, the crown of thorns, the nails, and the cross. The power of God is not manifested in creating the wrath; rather, it is in *enduring* the vessels of wrath "with much patience" in the personage of Jesus. Wrathful judgment is something the power of God endures; it is not something God sponsors.[3]

Why is human judgment wrathful? It is because people have repeatedly worshipped false gods to whom they have attributed their own desires for violence and scapegoating. This is why Jesus told his disciples, "The hour is coming when whoever kills you will think that he is offering service to God. And they will do this because they have not known the Father, nor me" (John 16:2–3). I think this is the reason that Christianity has a long and sad history of scapegoating violence: Christians have repeatedly made the error of projecting human wrath onto God. It is tempting to believe that God hates the same people we do, but projecting our own feelings and desires onto God is idolatry.

Why have translators of Paul's letter to the Romans so often converted the Greek orgé, "wrath" to "wrath of God," or "God's wrath"?

Most likely, translators have assumed that this is what Paul meant. Despite Jesus' teachings of love and forgiveness, people have always been powerfully drawn to self-righteous vengeance, and it has seemed natural and obvious to attribute such wrath to God. In addition, the Hebrew Scriptures have passages that seem to describe God as wrathful, perhaps because the ancient Hebrews found it impossible to believe that God did not seek retributive justice against evildoers. However, as discussed above, Deuteronomy 32:35 describes vengeance as the province of God, who can then choose whether or not to punish wrongdoing.

One can still believe that God makes judgments while attributing "the wrath" to humanity. Christianity teaches that God cares about all of creation, and it follows that God would likely condemn human destructiveness. Indeed, the Bible relates that God flooded the entire earth because "the earth was corrupt in God's sight, and the earth was filled with violence" (Genesis 6:11). Did the Flood demonstrate God's wrath? Perhaps the Flood was the only way God could stop violence. I see no necessary contradiction in holding that God is purely loving, and that God makes judgments, as long as God refrains from wrathful punishment. Perhaps such a God would allow humans to suffer the consequences of their own sinfulness, but I think such a God would prefer to reveal human sinfulness in ways that would encourage people to change their ways. I think that the life and death of Jesus encourages the latter approach.

The Faith of Christ

Many Christians believe that being justified, or being right with God, depends on whether or not Jesus has been accepted as Lord and Savior. In other words, faith, not works, justifies us in the eyes of God. This view largely derives from reading Romans 3:21–22 as follows: "But now the righteousness of God has been manifested apart from law, although the law and the prophets bear witness to it, the righteousness of God through faith in Jesus Christ for all who believe." However, the phrase "faith *in* Christ" could also be translated as "faith *of* Christ," with profound theological implications.[4]

The Greek phrase here is *pisteos Christou*, which is the genitive construction. It could be translated as either "faith in Christ" or "faith of Christ." In Romans 4:16, Paul implicitly used the genitive construction to describe the faith *of* Abraham. Obviously, he meant the faith of Abraham rather than faith in Abraham, because neither the Hebrews nor anyone else regarded Abraham as divine. When Paul clearly wished to communicate "in," he used the Greek word *en*. In Ephesians 1:15 and Colossians 1:4, *en* is used for faith *in* Christ, but neither passage states that faith in Christ is essential for justification. Further, scholars have doubts about Paul's authorship of these two epistles. Therefore, even though many English Bibles have *pisteos Christou* translated as faith *in* Christ, in Paul's undisputedly authentic letters, specifically Romans 3:22 and 3:26, Galatians 2:16 and 3:22, and Philippians 3:9, faith of Christ seems more appropriate. A difficulty is that translators, in trying to determine what particular passages mean, invariably impose their own theology and beliefs onto the text. There is no way for translators to know with any certainty what the original writers meant to convey. Translators who have been convinced that the New Testament aims to equate Jesus with God might have been prompted, perhaps mistakenly, to translate *pisteos Christou* as "faith in Christ."[5]

Why is this important? For one thing, if faith in Christ alone justifies us, works seem unimportant. However, on closer inspection, this theology still requires one work—to have faith in Christ.[6] I find this theology problematic, because it is too easy for some people and too hard for others. It seems to me that many people who believe that their faith alone justifies them show little interest in serving God's creation or making meaningful self-sacrifices in service to others. This attitude accords well with contemporary American consumerism and narcissism, but its relative indifference to those who are poor, weak, or vulnerable strikes me as at odds with Jesus' ministry. Conversely, faith in Christ can be exceptionally hard for those who have experienced great loss or profound suffering. They often feel abandoned by God.

In contrast to the work of having faith in Christ, our gaining the faith *of* Christ happens by grace. We know from Jesus' life, teachings, and death that the faith of Christ involves love, compassion, and caring. When this faith abides in us, we can find it soothing and empowering. However, if we cannot find that faith in our hearts and minds, we are not necessarily bad, evil, or unjustified in God's eyes. I do not think that God judges people unfavorably if their only "fault" is that they are unable to believe in God's love, whether their lack of faith has resulted from deep wounds or from their reflections on whether or not belief in God's love is reasonable.

Through God's grace (Romans 5:2, 5:15; 2 Corinthians 4:15; Ephesians 4:7; Titus 2:11; Hebrews 2:9, 12:15; James 4:6; 1 Peter 4:10), we can become transformed by the faith of Christ. Christians experience this by becoming new creations in Christ, and their works naturally reflect having Christ's faith. Paul wrote, "Therefore, if anyone is in Christ, he is a new creation; the old has passed away, behold, the new has come" (2 Corinthians

5:17). Though we are not saved by works *per se*, our loving, compassionate deeds reflect our adopting the faith of Christ. Jesus said, "Believe me that I am in the Father and the Father in me; or else believe me for the sake of the works themselves" (John 14:11). With such faith, we abide in God and God abides in us (see John 14:20, 15:10; 1 John 2:24, 4:6).

The faith of Christ encourages us to serve God. This, I think, is why James focused on works:

> What does it profit, my brethren, if a man says he has faith
> but has not works? Can his faith save him? If a brother or sister
> is ill-clad and in lack of daily food, and one of you says to them,
> "Go in peace, be warmed and filled," without giving them the
> things needed for the body, what does it profit? So faith by itself,
> if it has no works, is dead. But some one will say, "You have faith
> and I have works." Show me your faith apart from your works,
> and I by works will show you my faith (James 2:14–18).

It is noteworthy that, aside from the salutation, the only other time James refers to Jesus Christ is to emphasize the faith of Christ: "My brethren, show no partiality as you hold the faith of our Lord Jesus Christ, the Lord of glory" (2:1).

Another important implication of focusing on the faith of Christ is that such faith is not an individual choice or event. Gaining this faith almost always involves communal participation, which is why we need church communities to develop and maintain our faith. It is through the collective faith of the church community that people express the faith of Christ, supporting and inspiring each other. At the Pentecost, the Holy Spirit was poured out on everyone present, creating the church. And although the vicissitudes of life may strengthen or weaken our own, individual faith *in* Christ, an important role of community is to support its members by virtue of its collective expression of the faith *of* Christ. When we manifest the faith of Christ in our works, we make it easier for other people to receive that faith, which helps them cope with difficult situations and encourages them to perform works of love that help others.

This is one reason that the Christian Vegetarian Association (CVA) is an important ministry. Many CVA members despair over the terrible plight of so many of God's creatures. However, as a community, we can become empowered by the faith of Christ to believe in God's love and to believe that our struggles, however vain they might sometimes appear, glorify God and, in the final analysis, matter.

Guided by the Faith of Christ

I see Christian discipleship as calling us to "be perfect, as your heavenly Father is perfect" (Matthew 5:48). As inherently mimetic creations, we need a human model to emulate in order for us to seek perfection. God told Peter, James, and John, "This is my beloved Son . . . listen to him" (Matthew 17:5; Mark 9:7; see Luke 9:35), and Jesus instructed his disciples: "Follow me." Being Christian involves, among other things, choosing Jesus as a model, because Jesus modeled his life on God's desires. This is why Jesus said,

> Truly, I say to you, the Son can do nothing of his own accord, but only what he sees the Father doing; for whatever he does, that the Son does likewise. For the Father loves the Son and shows him all that he himself is doing; and greater works than these will he show him (John 5:19–20).

Paul understood well the importance of modeling our desires on Jesus. He told the Philippians, "Do nothing from selfishness or conceit, but in humility count others better than yourselves. Let each of you look not only to his own interests, but also to the interests of others. Have this mind among yourselves, which you have in Christ Jesus" (2:3–5). Jesus sought only to serve God. The "mind . . . in Christ Jesus" focused on God's desires, and our minds should do likewise. With Jesus as our model, we have the tools to accomplish this.

Philippians continues:

> Though he was in the form of God, [Jesus] did not count equality with God a thing to be grasped, but emptied himself, taking the form of a servant, being born in the likeness of men. And being found in human form he humbled himself and became obedient unto death, even death on a cross (2:6–8).

Christians are called to be humble and to serve, not be served. Jesus dramatized this by washing his disciples' feet (John 13:5–11), and then he instructed his disciples:

> If I then, your Lord and Teacher, have washed your feet, you also ought to wash one another's feet. For I have given you an example, that you also should do as I have done to you. Truly,

> truly, I say to you, a servant is not greater than his master; nor is he who is sent greater than he who sent him (13:14–16).

Service to God can satisfy our need for self-esteem, because such work is inherently meaningful. When we assess our self-worth by comparing ourselves to our peers, we constantly struggle to outperform them in a never-ending contest. In this human world of rivalry and conflict, people often feel they can never get enough money, power, or prestige. In contrast, by serving God, we can gain a sense of self-worth that has no relationship to our standing among our fellow humans. Most importantly, with Jesus as our model, we cannot be in rivalry with our model. Jesus rejected Satan's three temptations and desired to do the will of God rather than satisfy his human desires for comfort, wealth, or power. Human mimetic rivalries encourage people to compete with each other to establish who is God's greatest servant, but Jesus rejected this mindset. Jesus asserted that even he would not be the greatest servant and that his followers will do even greater works than he (John 14:12).

Humans have always generated and maintained community by scapegoating innocent victims, but the faith of Christ encourages us to generate and maintain community through love. Christians generally regard the Bible as the blueprint for inspiration and guidance, but the Bible has been a stumbling block to faith for some.

Faith and the Bible

Among the first Christians, faith was a matter of experience. The disciples who had earlier abandoned Jesus had an experience at Pentecost that inspired them to spread the gospel. Then, the early converts to Christianity heard stories about Jesus, and the experience changed their lives. In today's world, experiences continue to be an important part of Christian faith; these include experiences in our communities, in nature, and in prayerful meditation. Belief that God has worked through Jesus and that God continues to work through the Holy Spirit means that Christians profess a faith in divine action within the world—the same divine action described by the Bible's stories.

For reasons discussed earlier, many people find it difficult or impossible to trust in the Bible as a complete and accurate description of God. Nevertheless, the belief that God works within history, a belief that appears to be universal among Christians, can make it possible for them to believe that God's work did inspire the Bible. All Christians can then agree that the

Bible provides insights into God's desires for creation. Consequently, a Christian does not need to believe that the Bible's stories are perfectly accurate historically to believe that they reveal truths about humanity, human community, and God. Further, those with different approaches to the Bible can find common ground in the Bible's depiction of Jesus as reflecting God's compassion, mercy, and love. The Bible teaches that all Christians who strive to live in love like Jesus can come together in a faith community.

I think that central to Christian faith is a belief in the power of love, compassion, and forgiveness to create peaceful, harmonious communities. Jesus' life and teachings made it possible for him to be the last victim. Nevertheless, Christian authorities and communities have participated in scapegoating, which I see as a consequence of Christians not fully embracing the faith of Christ.

Despite the Bible's limitations for providing literal truths, I am impressed that many biblical insights accord with science. However, we should keep in mind that the Bible's stories were written by and for people who had far less understanding of the physical and social sciences than we have today.

Living out Faith

On the road to Damascus, Jesus did not say to Saul, "Saul, Saul, why don't you believe in me?" Jesus said, "Saul, Saul, why do you persecute me?" (Acts 9:4). Saul had been blind to his own participating in the scapegoating process. He had consented to the stoning of the innocent Stephen (Acts 7:58-8:1). Paul's experience with the resurrected Jesus left him blind. When we can see, we readily mimic the values and beliefs of those around us. Unable to see, Saul could search inside his own mind and start to recognize how he had participated in unjust violence. He was so transformed that even his name changed.

What we need is the mind of Christ (1 Corinthians 2:16; Philippians 2:5), which Christianity teaches is gained through the Holy Spirit. I am not convinced that only Christians have access to the mind of Christ, because I believe that it is within God's power to send as many messengers as needed to reach the world's people. For Christians, one important aspect of having the mind of Christ is that it can transform them from the human tendency to scapegoat to having faith in God's love and expressing that love. I think this is a central component of the experience of being born again, which inspired Paul to spread the gospel among the Gentiles. Christians are likewise called to "Go therefore and make disciples of all nations" (Matthew 28:19; see also Mark 13:10; 16:15, Luke 24:47).

Jesus emphasized the importance of following him:

> I am the vine, you are the branches. He who abides in me, and I in him, he it is that bears much fruit, for apart from me you can do nothing. If a man does not abide in me, he is cast forth as a branch and withers; and the branches are gathered, thrown into the fire and burned (John 15:5–6).

Having the mind of Christ helps Christians resist the temptation to participate in "righteous" violence. Without the mind of Christ, Christians tend to focus their lives on mimetic rivalries and other conflicts, which do not bring them closer to God. Therefore, Paul wrote, "If we live by the Spirit, let us also walk by the Spirit. Let us have no self-conceit, no provoking of one another, no envy of one another" (Galatians 5:25–26).

Christian living, which should include showing love and defending victims, can be inconvenient, challenging, or risky. For example, those of us who are animal advocates will be called to renounce animal-based foods that we once enjoyed; we might be called to struggle socially and financially; and we will often find ourselves ostracized or even persecuted. Nevertheless, Jesus reassured his followers, "For my yoke is easy, and my burden is light" (Matthew 11:30). Perhaps the reason is that, despite the hardships of discipleship, following Jesus can provide a sense of purpose and an inner peace.

Religion and Faith

To my reading, the Gospels indicate that Jesus sought to reform Judaism to God's original intentions for the Chosen People; Jesus did not try to start a new religion. He said, "Think not that I have come to abolish the law and the prophets; I have come not to abolish them but to fulfil them" (Matthew 5:17). Jesus aimed to establish an age of righteousness, which is why he echoed Hosea 6:6 when he said, "I desire mercy and not sacrifice" (Matthew 9:13; 12:7). Though Jesus initially focused his ministry on Judaism, evidently he envisioned a time when disciples would take his message to all the nations. It appears that, by these means, he strove to save the world (Mark 16:15; John 12:47). This view accords with the prophecy of Isaiah:

It is too light a thing that you should be my servant to raise up the tribes of Jacob and to restore the preserved of Israel; I will give you as a light to the nations, that my salvation may reach to the end of the earth (Isaiah 49:6; see also 42:6).

What about those who remain skeptical about Christian claims regarding Jesus' divinity, the nature of God, or even God's existence? Many people, after carefully reviewing the evidence, remain doubtful about such core Christian beliefs. Some find meaning and truth in one of the many non-Christian faiths, and I have no quarrel with them as long as their faith engenders compassion and respect. Others reject all religions and hold that we live in an exclusively materialist, ultimately meaningless universe. I think such a view fails to explain consciousness adequately. Scientists, who have associated consciousness with certain parts of the brain, have predicted that further research will be able to fully explain consciousness in materialist terms. I am skeptical, in part because identifying the neurotransmitters and brain locations associated with consciousness does not explain how one has subjective experiences. In particular, I am doubtful that any scientific explanation will account for *my* consciousness—how I came to perceive myself existing in this part of the world at this point in time as a being with a sense of my unique, constant identity. My consciousness, as well as that of others, invites metaphysical, nonmaterial convictions. I associate these metaphysical beliefs with God.

Regardless of our religious outlook, we need convictions about the way the universe ought to be in order to make choices that we believe are moral and meaningful. As an act of faith, I believe there is a chance, perhaps even a good chance, that there is a higher power that Christians call *God* that does care about creation. I aim to live *as if* God were about love, because it is a reasonable hypothesis, and because I cannot imagine enduring life's struggles without such a faith to inspire me. As Nuechterlein has observed, faith that God is about love does not engender violence or imperialism,[7] whether it is valid or not.

"There Is Neither Jew nor Greek"

People generally establish relationships with boundaries defined by gender, family, clan, and nation. In Acts 10, Peter learned in a dream that these are human, not divine, distinctions. Peter had been taught not to eat with Gentiles, but upon reflecting on a dream, he concluded, "God has shown me that I should not call any man common or unclean" (Acts 10:28). The sacred, hierarchical order is based on exclusion and has its roots in the violence of the scapegoating process. The sacred order heralded by Jesus is nonviolent, inclusive, and devoid of scapegoating victims. James Alison has written,

> Every local culture builds frontiers by means of victims; it is only if we begin from the forgiving victim [Jesus, who forgave those who abused and abandoned him] that we can build a culture which has no frontiers—we no longer have to build any order, security, or identity *over* [and] *against some excluded person.*"[8]

This is the critically important lesson that Acts 10 teaches; unfortunately, many Christians think the chapter's main message is that we may eat animal flesh. Although Peter dreamt that God told him that no food is inherently unclean, the biblical account then relates that Peter initially did not understand what the dream meant (Acts 10:17). Peter then concluded that the point of the dream was that Peter should eat with Gentiles. He did not conclude that killing and eating animals raises no ethical issues.

Paul emphasizes that Jesus sought to eliminate the boundaries that keep us from loving each other: "There is neither Jew nor Greek, there is neither slave nor free, there is neither male nor female; for you are all one in Christ Jesus" (Galatians 3:28). What about the distinction between humans and nonhumans? Are humans and nonhumans "one in Christ Jesus"? I think in one sense the answer is yes and in another it is no. Animals and humans have feelings and have the capacity to make moral choices (see Chapter 1), and they both are important parts of God's creation. However, according to the Bible, only humans are created in the image of God (Genesis 1:26), which gives humans special responsibilities. God instructs Adam to till and keep the Garden of Eden (Genesis 2:15), and it follows that Adam's descendents should similarly care for God's creation. Just as Jesus was called to serve God and God's creation, the degree to which we manifest our divine nature is the degree to which we answer the call to serve God.

Humans are distinctive, but not unique, in our focus on self-esteem, which tends to make us desire vengeance when we feel offended. Consequently, communities bounded by love need to be willing to forgive, a topic we shall turn to next.

Chapter 7: Forgiveness

Forgiveness versus Retaining Sins

After the Resurrection, Jesus greeted his disciples, "Peace be with you. As the Father has sent me, even so I send you . . . Receive the Holy Spirit. If you forgive the sins of any, they are forgiven; if you retain the sins of any, they are retained" (John 20:21–23). The disciples had abandoned Jesus in his time of crisis, even though Peter had promised never to abandon him (Matthew 26:33, 35; Mark 14:29; Luke 22:33; John 13:37). Yet upon his return, Jesus did not condemn the disciples. Rather, he said, "Peace be with you," which demonstrated that Jesus still loved his disciples and that he forgave them.

The disciples' experience of being forgiven for their betrayal helped prepare them to teach Jesus' ministry of love. Most people feel entitled, or obliged, to avenge perceived offenses against them. However, on finding that Jesus had, out of love, forgiven them, they could appreciate the power of love and the appropriateness of forgiveness. Similarly, when Jesus met Saul on the road to Damascus, he did not rebuke Saul; instead, Jesus said that he would tell Saul what to do. Calling Saul to discipleship was a form of forgiveness, and Saul was so transformed by the entire experience that he subsequently became Paul: Saul, the persecutor of Christians, believed his violence was righteous. Paul, forgiven for his grievous misdeeds, recognized his past errors and was ready for discipleship.

What did Jesus mean when he said, "If you forgive the sins of any, they are forgiven"? I think this means that forgiving sins allows people who believe they have been offended to move past old resentments.[1] Commonly, anger and resentment from perceived offenses remain in a person's heart; this causes misery to the person carrying these resentments, and it poisons the possibility for reconciliation. Therefore, "If you retain the sins of any, they are retained" means those who remain resentful and vengeful are unable to forgive and heal broken relationships.

After the Resurrection and immediately before departing from the disciples, Jesus said, "Thus it is written, that the Christ should suffer and on the third day rise from the dead, and that repentance and forgiveness of sins should be preached in his name to all nations" (Luke 24:46–47). Repentance and forgiveness of sins would be the disciples' principal teaching, because these are essential to establishing peaceful communities based on love.

Our desire to feel justified by God strongly encourages us to rationalize even our most violent or hurtful acts as righteous, and Proverbs says, "Every way of a man is right in his own eyes" (21:2). Christian faith teaches that God offers forgiveness to everyone (see, for example, 1 John 1:9), so we do not need to prove ourselves worthy of God's love. Relieved of the burden of justifying ourselves, we can face our sins and repent of our misdeeds. Still, we must choose to repent in order to *experience* God's forgiveness. Once we have experienced forgiveness, as the disciples and Paul experienced Jesus' forgiveness, we can become prepared to forgive those who have hurt us.

Forgiveness and Peace

Jesus tried to teach people, who naturally tend to fall into divisive rivalries, how to live peacefully with each other. He recognized the importance of forgiveness, and when Peter asked if he should forgive his brother as many as seven times, Jesus replied, "I do not say to you seven times, but seventy times seven" (Matthew 18:22). If we repeatedly forgive our brother, eventually he will very likely cease to offend us, because we will have given him no new cause to feel offended. If we act out against him, he will likely offend us again in "righteous" anger.

When expressing forgiveness, one should not convey an implicit accusation. If our words or actions communicate, "I am so magnanimous that I forgive you, even though you are a scoundrel," we convey a condescending, not a forgiving, attitude. We should acknowledge our own contribution to the conflict, ask for forgiveness, and then, if appropriate, express how we forgive the other. When our communication is respectful, loving, and compassionate, others can recognize ways in which they have been selfish or thoughtless. They may then adopt a more loving frame of mind, forgive us for our sins against them, and work with us toward reconciliation. If we are accusatory and judgmental, we become locked into conflicts that are ultimately resolved by physical or emotional violence. This dynamic helps account for numerous conflicts in families, in churches, and between nations that have undermined community and frustrated God's desire for the reconciliation of creation. Christians are called to be one body in Christ

(Romans 12:5; Galatians 3:28), but judging people excludes and divides. Therefore, Paul wrote, "Then let us no more pass judgment on one another, but rather decide never to put a stumbling block or hindrance in the way of a brother" (Romans 14:13).

Of course, even when our forgiveness is unconditional, it is not always accepted. Some people refuse forgiveness, either because they do not believe they have done anything wrong or because they cherish the resentments that forgiveness threatens to disarm. For many people, their resentments allow them to maintain a self-image as a victim, which excuses their anger and hostility. In these situations, we can only forgive in our own hearts and pray that the Holy Spirit will soften the hearts of others. We are to follow the instruction: "Put on then, as God's chosen ones, holy and beloved, compassion, kindness, lowliness, meekness, and patience, forbearing one another and, if one has a complaint against another, forgiving each other; as the Lord has forgiven you, so you also must forgive" (Colossians 3:12–13).

Anger versus Love

We tend to avoid angry, bitter people, because they generally make unpleasant company—and they can be dangerous. It seems to me that frequently, and perhaps always, such people have been deeply wounded. Often, their anger reflects a fear of being hurt again, and they express anger to keep people at a distance. If we are willing to patiently listen to their stories with an empathetic ear, withhold judgment, and have compassion for their pain, we will often ease their fear, lighten their burden, and reduce their anger.

Some people seem incapable of love, and this is likely because they have had little or no experience of being loved themselves. Over time, many such people come to feel ashamed and unworthy of love. Christian faith teaches that God loves everyone and offers forgiveness for all sins. However, many people have trouble believing that God loves them, because they know that they have sinned in serious ways and they do not feel forgivable. This is particularly the case if they, in my opinion mistakenly, regard God as wrathful (see Chapter 6). One possible reason that many Christians see God as wrathful is that they have been taught to regard God as "Father." They might project onto God their experiences with their own fathers, who might have been harsh, judgmental, and wrathful. However, seeing God as loving favors a view of God who resembles a loving parent, who might be disappointed in the child's behavior, but whose love for the child never wanes.

How can we be sure that God forgives us? Science can describe chromosomes, cells, and organs, but it cannot explain the spark of life; and Christians believe that the spark of all life comes from God. If God creates all life, it would be reasonable to believe that God loves all living beings, and an expression of that love would include offering us forgiveness for our transgressions. If we believe that we are loved and forgiven, we may forgive all who have wronged us, love everything, and enjoy the serenity that comes from following Jesus in obedience to God's will. This is the peace of mind enjoyed by the saints and by Jesus, who genuinely suffered on the cross but was still able to observe, presumably with equanimity, "It is finished" (John 19:30).[2]

Those who have been deeply wounded can have difficulty feeling loved by God. They are often angry at God for allowing their misfortune, and they find it hard to believe that God actually loves them if God permitted them to suffer so grievously. I think experiencing God's love is a form of grace. We can help people find and experience that grace by expressing God's love in our lives. To the degree that we make choices that hurt any part of God's creation, we show hardness of heart, and we fail to reflect God's love. Consequently, if we manifest hardness of heart, we impair our ability to fulfill our calling to preach the gospel to the nations (Mark 13:10; Luke 24:47). Hurtful actions do more than directly harm the victims; they tell the world, including those desperately in need of healing, that either we do not think it is important to reflect God's love, or that we believe that God is not loving.

Forgiveness and Judgment

It is natural for people to judge each other, but Jesus said, "Why do you see the speck that is in your brother's eye, but do not notice the log that is in your own eye?" (Matthew 7:3; Luke 6:41). The problem is that our judgments are always scandalous; we rarely see our own faults, and when we accuse other people, they defensively avoid seeing their own misdeeds. Consequently, judgments nearly always evoke resentment and hostility that, over time, can divide communities or lead to violence.

God does not participate in our judgments. Jesus taught that God "makes his sun rise on the evil and on the good" (Matthew 5:45). Indeed, Jesus asked God to forgive those responsible for murdering him "for they know not what they do" (Luke 23:34). Not passing judgment is different from discerning right from wrong. As the murderous mob descended upon St. Stephen, "He knelt down and cried out in a loud voice, 'Lord, do not hold this sin against them'" (Acts 7:60). Stephen identified their act as a sin, but evidently he did not judge them as evil or believe that they deserved severe punishment.

Should we forgive criminals? We should not aim to punish on the grounds of righteous vengeance, because our biases, prejudices, and tendencies to join the scapegoating mob render us ill equipped to determine who deserves punishment. However, a community's legitimate safety concerns might mandate imprisonment or other means to prevent people from harming others. Further, many believe that threat of punishment deters criminal behavior, and others believe that imprisonment can reform criminals. Given the human predilection for vengeance, it is difficult to know whether or not our punishments are grounded in compassion, concern, and respect for both the criminal and society at large.

In attempting to make such discernments, it is helpful to remain mindful that moral judgments do not require that we judge others as either saintly or contemptible. Jesus said, "Judge not, that you be not judged. For with the judgment you pronounce you will be judged, and the measure you give will be the measure you get" (Matthew 7:1–2; see also Luke 6:37). In other words, when we judge other people, we are liable to receive similar judgment. For example, if meat eaters feel condemned by vegetarians, they often become defensive, make reciprocal accusations of the vegetarians, and close their hearts and minds to the vegetarian message. On the other hand, when vegetarians denounce the institution of factory farming as cruel, they are discerning right from wrong practices. They are not judging the moral fiber of those involved in factory farming. If vegetarians refrain from attacking people or assuming a holier-than-thou posture, meat eaters are more likely to see vegetarianism as a loving and compassionate choice.[3]

In addition to impairing community-building, our judgments are often based on the false premise that *other* people, not we, deserve condemnation. In truth, we often contribute significantly to conflicts. When we judge other people, our judgment often constitutes scapegoating because it almost always involves projecting our guilt onto other people. Indeed, Paul wrote, "Therefore, you have no excuse, O man, whoever you are, when you judge another; for in passing judgment upon him you condemn yourself, because you, the judge, are doing the very same things" (Romans 2:1). When we fail to forgive, our position is often like that of the ungrateful debtor whose king forgave his large debt, but he then refused to forgive another man a much smaller debt (Matthew 18:23–34).

"Forgive Us Our Debts"

The Lord's Prayer includes, "Forgive us our debts, As we also have forgiven our debtors" (Matthew 6:12). Because we all sin, our only hope for favorable judgment from God is God's forgiveness. Despite our shortcomings, God offers forgiveness because "God sent the Son into the world, not to condemn the world, but that the world might be saved through him" (John 3:16–17). God offers us forgiveness, and consequently we should offer forgiveness to others.

Does God forgive unrepentant sinners? Many Christians think not. Even if God did forgive the unrepentant, such sinners would, by definition, refuse God's forgiveness. Consequently, to the degree that unrepentant sinners are not forgiven—either by God's withholding forgiveness or by their refusing it—they would distance themselves from God's love.

How can the Son save the world? From an anthropological perspective, Jesus differs from the heroes of primal religions who engage in "righteous" violence. These heroes vanquish the forces of evil and permit the weak and downtrodden to gain their rightful places of power. The problem is that the formerly weak and downtrodden quickly become the victimizing powers and principalities themselves. Undoubtedly, institutional Christian– ity has victimized many people, but these activities have distorted and misrepresented Jesus' ministry. Jesus taught that the only way to bring everlasting peace is through love, not vengeance. Therefore, the Son saves the world not by the power of the sword but by the power of love.

I think that Jesus was trying to teach that avenging perceived offenses is the wrong way to respond to injured self-esteem. Rather, the first step in gaining and maintaining self-esteem in a harsh and judgmental world is to recognize that others should not determine our worth. Our sense of worth and accomplishment should come from a conviction that God loves us and from participating with God in the reconciliation of all creation, however imperfectly we might perform this task.

How to Find Forgiveness in Our Hearts

While suffering and dying on the cross, Jesus said, "Father, forgive them, for they know not what they do" (Luke 23:34). This likely surprised the mob, and it has profound implications for Christian faith. First, note that Jesus asked God to forgive them; he did not announce that he forgave his tormentors himself. If Jesus had forgiven them himself, the mob would likely have rejected his forgiveness, because the mob felt entitled to its actions and did not believe it needed forgiveness.

Jesus asked God to forgive them. When one genuinely loves everybody and everything, one wants them to be forgiven, even for their crimes. One will want evildoers to desist from hurting other individuals, of course, but love vanquishes one's desire for vengeance. Being human, we find it hard to forgive those who have deeply wounded us. When we find it impossible to forgive, sometimes the best we can do is, like Jesus, to pray for God to forgive them, just as God offers forgiveness to sinners like us.[4]

Great spiritual leaders have offered guidance on how to find forgiveness in our hearts. Borrowing from a wide range of sources, I offer the following thoughts that have been helpful for me. In deep prayer or meditation, I first reflect on God and on my belief that, as a creation of God, I am loved and forgiven by God. This helps give me the strength I need to acknowledge my weaknesses and sins. Then, I reflect on my anger: Why am I angry? How is my anger coloring my life? Am I mindful of the pain, suffering, and conflicts in the lives of those toward whom I am angry? Then, I reflect on my guilt: Why do I feel guilty? What can I do to atone for the harmful things I have done? Then, I reflect on my shame: When did I feel ashamed? Do I think God was disappointed in me? What have I done of which God would disapprove? Facing the inner voices of anger, guilt, and shame that haunt me, rather than trying to repress those voices, I find it much easier to forgive other people and to forgive myself. Such forgiveness helps provide a sense of inner peace that makes it much easier for me to express love and compassion.

A Lesson in Forgiveness: The Adulteress

The story about the adulteress (John 8:3–11) relates that the scribes and Pharisees tried to trap Jesus by asking whether they, acting according to the law of Moses, should stone the guilty woman. Initially, Jesus did not reply; instead, he wrote something in the sand. This broke the mob's momentum toward stoning her. If they had not been forced to pause and think, they would have stoned her regardless of Jesus' response. Then, Jesus wrote in the sand once again after saying, "Let him who is without sin among you be the first to throw a stone at her" (John 8:7). Why did Jesus write in the sand? Mimetic theory suggests that, if Jesus had met the mob's gaze, the angry accusers would have projected their anger onto him.[5] They would likely have regarded Jesus' gaze, however loving and compassionate it was, as an accusation and an affront; and they might have killed Jesus, as well as the adulteress.

When Jesus challenged the crowd to produce someone without sin to cast the first stone, he was demanding that someone step away from the crowd and take responsibility for the violence. Mimetic theory posits that people are very reluctant to act without prompting from someone else, and indeed nobody came forward to commence the stoning.

Jesus forgave the adulteress her sin before she asked for forgiveness or even expressed repentance or regret; then, he told her to sin no more. If Jesus had demanded her repentance, she would likely have sought excuses for her behavior, because presumably she had once felt justified in committing adultery. Jesus showed that her sin was forgivable. Therefore, she did not need to find excuses for her behavior, and she could then repent of her sin and resolve not to commit it again.

Born Again

Jesus said, "Truly, truly, I say to you, unless one is born anew, he cannot see the kingdom of God . . . unless one is born of water and the Spirit, he cannot enter the kingdom of God" (John 3:3, 5). Rebirth requires water, a universal symbol of renewal that mixes everything together and washes away differences. Without differences, we no longer see ourselves as better or worse than our neighbors, and this allows us to be aligned with Christ. Rebirth, Jesus noted, requires the intercession of the Spirit, because human desires encourage us to sin and call it righteousness.

If one is born again, part of what dies is the human sin of believing that God ordains one's "righteous" violence. We can see God as loving and forgiving and become prepared, as Jesus instructed, to "love your neighbor as yourself" (Matthew 22:39; Mark 12:31; see Luke 10:27). Being born again means becoming "in Christ," and Paul wrote, "Therefore, if any one is in Christ, he is a new creation; the old has passed away, behold, the new has come" (2 Corinthians 5:17). Nevertheless, even after experiencing such a spiritual transformation, it takes great faith to believe that the path to peace is not by force but rather by love and forgiveness.

For Christians, being born again is a matter of choosing Jesus as the center of our faith and the model for our actions. Can people of other faiths be born again? I think that anyone can experience rebirth if they have a sense of transformation that compels them to seek to renounce acquisitive mimetic desires and to adopt God's[6] loving desires for all creation. What if someone has not undergone the ritual of baptism? I think this ritual was understood by Jesus' audience to symbolize the cleansing of sins, but there are different rituals by which people in other cultures can experience rebirth.

Given that most humans lived before Jesus' ministry, and that many people have little or no exposure to Christianity today, I am doubtful that God reserves rebirth only for Christians. I think rebirth has always been available to anyone who is open to the grace of God.

Does being born again mean that one no longer sins? Paul lamented, "For I do not do the good I want, but the evil I do not want is what I do" (Romans 7:19). Though Paul was evidently discouraged by his shortcomings, he acted in a spirit of repentance. Born again, Paul desired to sin no more. Though we always fall far short of God's perfection (Romans 3:23), being reborn helps us align our desires with God's desires. The prophet Jeremiah prophesied how this might be possible:

> This is the covenant which I will make with the house of Israel after those days, says the Lord: I will put my law within them, and I will write it upon their hearts; and I will be their God, and they shall be my people. And no longer shall each man teach his neighbor and each his brother, saying, 'Know the Lord,' for they shall all know me, from the least of them to the greatest, says the Lord; for I will forgive their iniquity, and I will remember their sin no more (Jeremiah 31:33–34).

When the Hebrews left Egypt, they needed the law to keep order and avoid anarchy. Jeremiah foresaw of a time when the Hebrews would no longer need the law. Instead, there would be a new covenant between God and God's people, in which God's law would be written on everyone's heart, and nobody would need instruction. The law would be known by the least and greatest. Isaiah similarly prophesied a time in which "the earth shall be full of the knowledge of the Lord," and all creation will live harmoniously and nonviolently (Isaiah 11:6–9, 65:25).

The Sunflower

Christians are called to love and forgive their enemies, even those who have abused or continue to abuse them. Can we forgive on behalf of other individuals who have suffered and continue to suffer at human hands? Simon Wiesenthal's remarkable story *The Sunflower* explores this question.[7]

While in a Nazi concentration camp, Wiesenthal was called to the bedside of a dying German soldier, who confessed to participating in the murder of about 300 Jews. They had been crowded into a building that was set ablaze, and he and fellow soldiers shot those who tried to escape out the windows. The soldier asked Wiesenthal, a Jew, to forgive him. Wiesenthal listened to the soldier's entire story, allowed the soldier to take his hand, and then left without speaking. The soldier died the next day and left all his possessions to Wiesenthal, who refused them. Wiesenthal caused considerable consternation among his friends in the concentration camp when he asked them whether he had done the right thing in allowing the Nazi to tell his story and request forgiveness. They could not understand how Wiesenthal could have any concern or compassion for a Nazi, because Nazis had murdered their relatives and friends and would likely murder them.

Having miraculously survived the concentration camp, after the war Wiesenthal asked dozens of people from a wide range of backgrounds whether or not he had done the right thing, and their varied responses represent the bulk of *The Sunflower*. Why did Wiesenthal want to know whether he had acted rightly? I think he needed to know whether the Nazis had destroyed his beliefs and values. They had killed many members of his family, stolen his possessions, and reduced him to a pathetic, starving, miserable man. Had they destroyed his faith that God was on the side of the good? Had they taken away his ability to respect all life, which derives from God? Could they make him curse God, as Satan had predicted Job would curse God (Job 1:11)?

Wiesenthal could not forgive on behalf of people he never knew, but his listening to the dying man's story showed compassion and silently communicated that God might forgive the soldier. Does forgiveness mean that there should be no consequences for destructive behavior? Wiesenthal did not think so. Although he showed human compassion by listening to the dying Nazi soldier, after the war Wiesenthal was a leading figure in efforts to capture Nazi criminals. He believed it was necessary to bring them to justice, so that future generations would know that people must be held accountable for their actions.

Should we seek to punish those who perpetrate heinous crimes, as Wiesenthal sought to do? Nearly everyone can empathize with his desire to bring Nazi criminals "to justice." In a world in which many people are not filled with the Holy Spirit, fear of punishment discourages criminal behavior. Nevertheless, as we deliberate on how to respond to crime and other anti-social behavior, I think Christian faith encourages us to focus on preventing harm to innocent individuals rather than on punishing criminals. We should aim to find compassion in our hearts for both victims and victimizers, particularly because so few of us are totally innocent or totally depraved.

One difficulty with Wiesenthal's quest for "justice" is that those who have committed offenses often perpetrate further crimes to avoid prosecution. Gangs often require initiates to commit a crime, because it is difficult to leave a gang that knows about one's illegal activities. The Truth and Reconciliation Commission (TRC) in South Africa illustrates an approach that reduces the likelihood that the quest for justice will escalate violence. The Commission invited victims of Apartheid to relate their stories, and many of those who perpetrated political crimes were offered amnesty—if their crimes were not excessively heinous, and if they fully disclosed their crimes. The TRC sought to procure justice through apology, forgiveness, and restitution. The process did not satisfy everyone, but it did facilitate a peaceful transfer of power from the white minority to majority rule.[8]

I think that Wiesenthal, by listening to the dying Nazi soldier's story, demonstrated that Nazi crimes against him and his family had not vanquished his ability to show compassion in the face of evil. Can the power of love overcome evil?

Chapter 8: The Power of Love versus the Power of Satan

God Is Love

The writer of 1 John declared, "He who does not love does not know God; for God is love" (1 John 4:8). If God is love, then God does not embrace cruelty, callousness, heartlessness, or vengeance. Indeed, "This is the message we have heard from him and proclaim to you, that God is light and in him is no darkness at all" (1 John 1:5).

Evidently, this writer recognized that, without love, we fall into bitter rivalries that often lead to violence. He wrote, "He who does not love remains in death. Any one who hates his brother is a murderer, and you know that no murderer has eternal life abiding in him. . . . Little children, let us not love in word or speech but in deed and in truth" (1 John 3:14–15, 18). Words alone are meaningless; love involves action. Paul expressed similar sentiments when he wrote, "Love does no wrong to a neighbor; therefore love is the fulfilling of the law" (Romans 13:10). These writings accord with Jesus' comment, "Not every one who says to me, 'Lord, Lord,' shall enter the kingdom of heaven, but he who does the will of my Father who is in heaven" (Matthew 7:21). Relating this to animal issues, many people say they "love animals," but truly showing love means helping animals in need and doing one's best not to harm them. Throughout his ministry, Jesus taught that God wants us to love:

> A lawyer stood up to put him to the test, saying, "Teacher, what shall I do to inherit eternal life?" He said to him, "What is written in the law? How do you read?" And he answered, "You shall love the Lord your God with all your heart, and with all your soul, and with all your strength, and with all your mind; and your neighbor as yourself." And he said to him, "You have answered right; do this, and you will live" (Luke 10:25–28).

Note that the first instruction is to love God totally; then, love your neighbor as yourself. If a person loves God, it follows that the person should show respect for God by caring for God's creation, which includes fellow humans, God's animals, and God's earth.

The lawyer then asked who is a "neighbor," and Jesus replied with the parable of the Good Samaritan. In this parable, and in many other biblical passages, love involves action, which accords with Jesus' saying, "Do this, and you will live." Showing love, rather than just claiming to love, is how we discern true prophets from false prophets. Jesus said, "Beware of false prophets, who come to you in sheep's clothing but inwardly are ravenous wolves. You will know them by their fruits" (Matthew 7:15–16). Along these lines, John wrote, "If any one says, 'I love God,' and hates his brother, he is a liar" (1 John 4:20); and Paul wrote, "The fruit of the Spirit is love, joy, peace, patience, kindness, goodness, faithfulness, gentleness, self-control; against such there is no law" (Galatians 5:22–23).

The Gospels relate only one commandment from Jesus' mouth: "A new commandment I give to you, that you love one another; even as I have loved you, that you also love one another. By this all men will know that you are my disciples, if you have love for one another" (John 13:34–36). Paul offers similar instruction:

> Owe no one anything, except to love one another; for he who
> loves his neighbor has fulfilled the law. The commandments,
> "You shall not commit adultery, You shall not kill, You shall not
> steal, You shall not covet," and any other commandment, are
> summed up in this sentence, "You shall love your neighbor as
> yourself" (Romans 13:8–9).

If we love our neighbor as ourselves, we will desire their well-being as much as we desire our own, and divisive mimetic rivalries will vanish.

Agape

Jesus' encounter with Peter after the Resurrection illustrates a loving interaction. Three times, Jesus asked Peter whether Peter loved him. The first two times that Jesus queried Peter, Jesus used the verb corresponding to the Greek word *agape* that describes a total, unconditional, universal love.[1] The first two times, Peter responded affirmatively, using the verb corresponding to the Greek *philia* that communicates a more restricted concept of brotherly love. The third time, Jesus utilized the verb *phileo*, indicating that he loved and forgave Peter so much that he was even willing to accept Peter's understanding of love (John 21:15–17).[2]

119

Though Peter was exasperated by Jesus asking him three times, this allowed Peter to undo his earlier denial of Jesus three times. Previously, Peter believed that he was so devoted that he would never betray Jesus. He had denied that he would ever abandon Jesus (Matthew 26:35; Mark 14:31; Luke 22:33), yet he did just that in Jesus' hour of need. Peter learned that he had participated in the crucifixion of Jesus, and we do likewise when we fail to show love, because Jesus said, "As you did it to one of the least of these my brethren you did it to me" (Matthew 25:40).[3]

I think God desires all our relationships to be grounded in a complete, unconditional, universal *agape*, and 1 John 4:8 and 4:16 reads "God is *agape*." Such love accords with God's desire for peace and harmony throughout creation. *Agape* is not grounded in the specific characteristics of other individuals. Regardless of any attributes or faults people might have, *agape* relates to our reverence for God and is based on the conviction that *God* loves everyone.

Covenantal and Committed Relationships

The Bible describes covenantal relationships as models for *agape*. Examples include God's covenant with all creation not to flood the earth again (Genesis 9); with humans and all creatures to one day establish peace and harmony (Hosea 2:18); with Abraham, Isaac, and Jacob to make them patriarchs of great nations (Genesis 17; Leviticus 26:42); and with the Hebrews to provide the Promised Land (Exodus 6:4) and the Ten Commandments (Exodus 34:28). All of these covenants were gifts, but in certain instances—such as with the Ten Commandments—the Hebrews were obliged to accept the gifts or suffer the consequences (i.e., dissolution of their community). In the New Testament, at the Last Supper, Jesus took the wine and said, "This is my blood of the covenant, which is poured out for many for the forgiveness of sins" (Matthew 26:28; see also Mark 14:24, Luke 22:20).[4] All of our relationships should model Jesus' covenant with his disciples, characterized by love, compassion, and forgiveness. This is why Jesus' commandment to his disciples was that they love one another (John 15:12). If we love each other as Jesus loved his disciples and as God loves all creation, we will seek respectful, compassionate, nonviolent solutions to conflicts.

Our personal, committed relationships should also be grounded on *agape*. Relationships based on acquisitive mimetic desire are often troubled. In the novelette *Kreutzer Sonata*, Leo Tolstoy shows the pitfalls of such love. Once the initial excitement of falling in love wears off, resentments build. The thrill of love invariably wanes, as does the sense of self-esteem that accompanies "winning" the affection of a desirable object of love; and differing goals and priorities cause discord. Once the object of love has been won, there is no further benefit to self-esteem. Indeed, as the object of love ages and becomes less and less a source of acquisitive mimetic desire, one may be inclined to seek sexual conquests among those deemed more desirable by one's peers.

Because love generated by acquisitive mimetic desire is related to self-esteem, it is difficult for such love to last. For example, Anne might find Bob attractive, in part because Bob expresses affection for Anne and vice versa. This raises mutual self-esteem, but it is a tenuous situation. For one thing, in the initial passion of love, one may be blind to the other's faults, which become unavoidably obvious eventually. As Anne comes to recognize Bob's faults, in terms of self-esteem, Anne finds Bob's high regard for her less satisfying. Furthermore, as Anne realizes that Bob knows of Anne's faults, Anne's regard for Bob's *opinion* will wane if Bob continues to express unqualified admiration. Consequently, it is difficult for love grounded in acquisitive mimetic desire, spurred by a quest for self-esteem, to sustain long-term relationships.

On the other hand, if we see God as the mediator of committed relationships, it is much easier to keep the promise to love and cherish each other "in sickness and health, until death do us part." One's primary commitment is to God. In other words, even though our life partners have faults, and even though our life partners might annoy, irritate, or even offend us, we are called to love and honor them, because we are committed to them through a divinely ordained relationship.

What about people who abuse their spouses or children? Should people stay in destructive or even dangerous relationships? I think physical or emotional abuse of a spouse or child violates the marriage covenant and releases the victims to seek protection, healing, and growth elsewhere.[5]

"For God So Loved the World"

John wrote, "For God so loved the world that he gave his only Son, that whoever believes in him should not perish but have eternal life" (John 3:16). If God were only interested in saving from death those who believe in the Son, it would have made no sense to mention that God gave the Son for the benefit of the *world*, including those who are intellectually unable to believe, such as young children, mentally disabled people, and animals. I think that Jesus came to reconcile all creation, not just those who have faith in Jesus. This is why "the creation waits with eager longing for the revealing of the sons of God" (Romans 8:19). Those "sons of God"[6] have faith in God's redemptive power, and they, being new creations in Christ, will herald a new age where "he will wipe away every tear from their eyes and death shall be no more" (Revelation 21:4).

How could God eradicate scapegoating and victimization? If God killed those who harm others, it would have been unfair, because those who engage in scapegoating violence largely "know not what they do" (Luke 23:34). Killing those priests who did not have conscious evil intent would have been tantamount to scapegoating. Most importantly, it would have been only a temporary solution to the problem of violence, because people would have persisted in a sacrificial mentality and would likely have concluded that the priests had been killed because they did the sacrifices incorrectly. Leviticus 10:1–2 describes the story in which Aaron's two sons are killed by God for offering "unholy fire." It is possible that a more accurate historical account was that their sacrifices did not relieve a sacrificial crisis, and *the people* killed Aaron's sons, believing that doing so satisfied the obvious wrath of an angry God. Interestingly, Leviticus 10:3 relates,

> Then Moses said to Aaron, "This is what the Lord has said, 'I will show myself holy among those who are near me, and before all the people I will be glorified.'" And Aaron held his peace.

Although this passage is open to a range of interpretations, perhaps Aaron recognized that, with everyone believing that God endorsed his sons' deaths, it was ill advised to resist the consensus.

What does "everyone who believes in him may not perish but may have eternal life" mean? Many people think "eternal life" refers to a permanent afterlife. I think that "eternal" communicates a different concept. "Eternal" means unbounded by time, which describes God's existence much better than human existence, because human lives inexorably head toward death. I think we can experience eternal life when we feel connected to the timeless universe. One way that Christians and many other people of faith can get this sense of connection to the universe is by aligning themselves with God, the creative force of the universe. We can feel aligned with God by serving God, for example by caring for God's creation.

We gain further insight from Buddhism, which I regard as a philosophy that does not necessarily compete with Christianity. The above notion of "eternal" accords with the state of mind that the Buddhists call "being awake" or "enlightenment," in which mindfulness is so complete that the awakened no longer feels trapped in a suffering and dying body but rather feels perfectly open to and accepting of the cosmos.

Do humans and animals have an afterlife? I do not think it is reasonable to believe that humans have an afterlife and animals do not. Humans and animals have much in common at genetic, physiological, and emotional levels. Some have argued that only humans have a soul, yet the Hebrew Scriptures use the same words, *nephesh chayah*, to describe the essence of both humans and animals. When relating to humans, translators of Genesis 2:7 have called *nephesh chayah* "soul" (King James Version) or "being" (Revised Standard Version). In Genesis 2:19, which refers to animals, they have translated *nephesh chayah* as "creature." Those who have used these verses to claim that only humans have souls have relied on translators' biases and not the Scriptures themselves. Indeed, the author of Ecclesiastes wrote,

> For the fate of the sons of men and the fate of beasts is the same; as one dies, so dies the other. They all have the same breath, and man has no advantage over the beasts; for all is vanity. All go to one place; all are from the dust, and all turn to dust again. Who knows whether the spirit of man goes upward and the spirit of the beast goes down to the earth? (Ecclesiastes 3:19–21).

Paul's writings indicate his belief in an afterlife. He wrote to the Philippians, "My desire is to depart and be with Christ, for that is far better. But to remain in the flesh is more necessary on your account" (Philippians 1:23-24; see also Galatians 2:20 and Romans 6:3–4). Similarly, Christian tradition holds that Jesus was resurrected from the dead and that an afterlife awaits everyone.

Many people who have had a "near-death experience" (NDE) relate a conscious, out-of-body experience that has convinced them of an afterlife.[7] However, it is difficult for those who have not had such experiences to evaluate NDE claims. Upon dying, I do not know what will happen to the stable sense of self that I carry throughout my life that seems unchanged even while most of my body's cells die and are replaced. Our fear of death, which relates to the destruction of the self, encourages us to envision some kind of existence after our bodies have ceased to function. Perhaps there is an afterlife, but regardless of the fate of the self, Christian faith offers the possibility of eternal life. This faith is grounded in a conviction that God is about life and love, not death and destruction. However, we often find that love is frustrated by "satanic" powers that seed discord, discontent, and death. What is the nature of Satan, and how should loving, compassionate people deal with satanic powers?

Satan

Many people regard Satan as a godlike figure. However, such a view lends itself to the scapegoating process, because people can believe that they are aligned with a good god while a satanic god possesses other people. Jesus' interactions with Satan indicate that Jesus did not regard Satan as a god. After rejecting Satan's three temptations (Matthew 4:1–11; Luke 4:1–13), Jesus cited scripture to Satan: "You shall worship the Lord your God." This passage identifies the Lord as God, indicating that Satan is not a god. If Satan is not a god, Satan must relate to humanity, in which case it is possible to understand Satan anthropologically, in terms of human culture.[8] And if we regard Satan anthropologically, we can see satanic violence and destructiveness as consequences of the human tendency to fall into mimetic rivalries.

We see satanic human desires in Mark's Gospel, where a central theme is that the disciples had great difficulty understanding Jesus' message. They repeatedly competed with each other for prominence, and consequently they often failed to recognize that God wants us to reflect God's love by serving each other. The disciples expected Jesus to become glorified and powerful, and they eagerly anticipated gaining power and prestige as Jesus' closest associates. Consider Mark 8:31–33 (see also Matthew 16:23):

> And he began to teach them that the Son of man must suffer
> many things, and be rejected by the elders and the chief priests
> and the scribes, and be killed, and after three days rise again.
> And he said this plainly. And Peter took him, and began to rebuke
> him. But turning and seeing his disciples, he rebuked Peter, and
> said, "Get behind me, Satan! For you are not on the side of God,
> but of men."

Jesus knew that his destiny included allowing himself to be rejected, killed, and raised again, which revealed the scandal of sacred violence and gave humanity an opportunity to reject the satanic tendency to scapegoat. Peter, however, had different ideas, hoping to gain power and glory as Jesus ascended to power. Peter's acquisitive mimetic desire was derived from seeing what other people wanted.

Satan resembles a transcendent, godlike figure *only* when people regard Satan as a divine individual with the power to force people to do things. For those who regard destructiveness as a consequence of universal *human* acquisitive desires or other human attributes, Satan has lost transcendence.[9] Therefore, Jesus said, "I saw Satan fall like lightning from heaven" (Luke 10:18). At best, satanic desires can tempt us, but they do not control us. When we lose faith in Satan's power, we are less inclined to accede to our own satanic desires, or to regard others as irremediably "possessed" individuals who endanger the community and must be expelled or killed.

What Is Satan?

The Bible has several passages that seem to describe Satan as a separate individual, particularly in the Book of Job, in which Satan converses with God. However, to my reading, the Book of Job is a parable that explores important philosophical and theological questions related to righteousness, faithfulness, and the problem of evil in a world governed by God. Therefore, when I discuss "Satan," I refer to the "satanic" part of the human psyche that is harmful and destructive.

There is ongoing conflict within the human soul between egocentric, self-serving desires, and desires to be loving and compassionate. When we find ourselves tempted to yield to illicit or harmful desires, we tend to blame other individuals for our shortcomings. Jesus squarely faced his own desires, and he found that he could overcome them. Commonly, people blame Satan for those harmful, destructive desires of which we are ashamed. I think we should recognize these desires as universal and part of what it means to be human. When we try to deny our own satanic desires, we nearly always proceed to project our satanic desires onto others and blame them for conflicts.

The Bible repeatedly describes God choosing people to serve as prophets or disciples, but Satan's attributes include being an accuser and a trickster. Satan's most effective trick is to convince people that Satan the accuser is God the chooser.[10] By dividing the world into good and evil, Satan convinces people that their satanic accusations represent God's choices. In the Book of Job, Satan orchestrates Job's travails, and Job's "friends" then assist Satan's work by falsely accusing Job of offending God. Like Job's friends, the satanic desire to make ourselves feel justified by God encourages us to accuse other people of being evil.

The mob that called for Jesus' crucifixion believed that Jesus was evil and that they were righteously abiding by God's will. As discussed in Chapter 3, the only way Jesus could unequivocally reveal Satan's trick of false accusation was to allow himself to be an innocent victim. This does not mean that we should be silent in the face of injustice. We are right to oppose harming any innocent individual, and I think Christianity encourages us to do this by identifying the *action* as sinful. However, if we accuse *people* of being evil, judging them as possessed by satanic motivations, we are playing the satanic game of accusing others in order to elevate ourselves and feel chosen.

Can Satan Cast Out Satan?

Mark reads:

> And the scribes who came down from Jerusalem said, "He is possessed by Beelzebub, and by the prince of demons he casts out the demons." And he called them to him, and said to them in parables, "How can Satan cast out Satan? If a kingdom is divided against itself, that kingdom cannot stand. And if a house is divided against itself, that house will not be able to stand. And if Satan has risen up against himself and is divided, he cannot stand, but is coming to an end" (Mark 3:22–26; see also Luke 11:14–18).

The human forces that have accused people of satanic possession and then tried to eradicate Satan from their midst have been satanic. "Satan casting out Satan" describes the scapegoating process.[11] When people call someone Satan, they have assumed the role of Satan the accuser, and invariably they come to participate in scapegoating. Jesus taught that the way of Satan is accusation, but the way of God is forgiveness, which breaks the cycle of accusation.

That Satan casts out Satan is true only in the sense that satanic forces do cast out the perceived Satan in a community's midst. However, attempts by Satan to cast out Satan merely divide the house and set the stage for future conflicts. Indeed, as long as scapegoating is the glue that holds communities together, the main difference between the victims and the victimizers is that the victimizers happen to have power at the moment. As long as Satan tries to cast out Satan, there will always be communal strife. The only way to break the endless cycle of violence is to develop a new culture grounded on love and forgiveness, which is what Jesus taught.

Inspired by Jesus and perhaps assisted by the Holy Spirit, we may refuse to participate in scapegoating. If necessary, we may choose to assume the role, like Jesus, of the willing and forgiving victim. Whether or not satanic violence is self-defeating, faithful Christians are called to imitate Jesus and participate in the reconciliation of creation by being peacemakers (see Chapter 10). Some Christians hold that we should willingly submit to, rather than resist, scapegoating.[12] Otherwise, the mob, unable to recognize its own satanic inclinations, will regard our resistance as satanic. Only if the mob recognizes the victims of scapegoating as innocent can the mob recognize that it is scapegoating; otherwise, it sees its own scapegoating as righteousness and justice.

I think this helps explain 1 Peter 2:18 and 3:1, which encourage slaves to obey their masters and wives to submit to their husbands. Importantly, the author of 1 Peter maintains that Christians, by their example of love, will encourage others to show love. Although love is an antidote to abusive relationships, this advice leaves people vulnerable to victimization until abusers adopt the faith of Christ. Will anyone defend victims?

The *Parakletos*: Helper and Defender

John 14:15 begins, "If you love me, you will keep my commandments." In other words, persons who loves Jesus will love God with all their heart, soul, mind, and strength, and their neighbor as themselves (Mark 12:30–31; Luke 10:27). John 14:16–17 continues,

> And I will pray the Father, and he will give you another
> Counselor [*parakletos*], to be with you for ever, even the Spirit
> of truth, whom the world cannot receive, because it neither sees
> him nor knows him; you know him, for he dwells with you,
> and will be in you.

The Greek word *parakletos* is commonly translated as "the one who defends the accused"[13] or "helper." To my reading, Jesus was a counselor who assisted those who were sick, poor, widowed, or otherwise disenfranchised, who were frequently victims of scapegoating. John described "another Counselor" as the "Spirit of Truth," in John 14:17, and as "the Holy Spirit" in John 14:26.

Jesus said that the counselor "dwells with you." How can we, as counselors, assist the accused? One way is to manifest the "Spirit of truth" by showing that the accusers have heaped excessive guilt upon the accused.

This is a hallmark of hypocrisy, making oneself seem righteous by condemning other individuals. So, in defending the woman accused of adultery (John 8:3–11), Jesus pointed out the hypocrisy of the accusers, because they felt entitled to pass judgment despite being sinful themselves (see also Matthew 23:35–37 and Mark 7:1–9). However, Jesus was but one person, and he recognized that the Holy Spirit would inspire future generations to assist and defend the accused. Therefore, Jesus said, "Truly, truly, I say to you, he who believes in me will also do the works that I do; and greater works than these will he do, because I go to the Father" (John 14:12).

Adding further insight to the workings of the Counselor, John 16:7–11 reads,

> It is to your advantage that I go away, for if I do not go away, the
> Counselor will not come to you; but if I go, I will send him to
> you. And when he comes, he will convince the world concerning
> sin and righteousness and judgment: concerning sin, because they
> do not believe in me; concerning righteousness, because I go to
> the Father, and you will see me no more; concerning judgment,
> because the ruler of this world is judged.

Jesus said that the Counselor will teach the world about sin, righteousness, and judgment. I will offer a Girardian perspective:[14] Concerning sin, the Holy Spirit will reveal that human communities have not believed in Jesus' way of love, compassion, and peace. They have blamed scapegoats rather than accept responsibility for their own sins.

Concerning righteousness, the Holy Spirit will reveal that the sense of righteousness that always accompanies scapegoating is mistaken. Jesus, judged sinful by the mob, was resurrected and went to the Father, thus proving that people have been wrong about righteousness.

Concerning judgment, the Holy Spirit will judge "the ruler of this world," and the ruler of this world of violence has been Satan, the accuser and trickster. The Holy Spirit will judge Satan as wrong, because God's righteousness is about love, not vengeance.

Faith in God's Love versus Fear of Death

Even though Christianity teaches that God is about love, we often have difficulty appreciating God's love, because we tend to be obsessed with death. As discussed in Chapter 1, all human cultures have needed to address death-related anxieties. Can belief in Jesus' resurrection and the eventual resurrection of our souls to a state of everlasting bliss mitigate the harmful social consequences of our fear of death? In other words, can faith in universal resurrection neutralize our fear of death, reduce our need to continually prove our self-worth, and position us to be more loving and compassionate? I think this can happen, though this would be difficult if we regarded God as wrathful and vengeful. If we saw God as wrathful, then we would likely hold that God restricts everlasting bliss in heaven to certain people and relegates those whom God despises to either everlasting nothingness or everlasting misery in hell. Belief in a wrathful God would encourage us to focus on the question of who will receive God's heavenly reward. As we sought to determine who "deserves" to go to heaven, we would desperately want to believe that we will be among the elect. In our attempt to feel chosen, we would seek reasons that others are not chosen. This mindset would tend to engender conflicts with people whose personal behavior, religious rituals, or religious beliefs differ from ours. Furthermore, if alleviation of our mortality anxieties required believing a specific set of religious tenets, we would tend to resist, with violence if necessary, other belief systems that contradicted our own.

Jesus did not avoid death; he died just as all of us will die. Jesus demonstrated that we can vanquish the *power* of death to rule our lives. Paul articulated this well in his first letter to the Corinthians:

> For the word of the cross is folly to those who are perishing, but to us who are being saved it is the power of God. For it is written, "I will destroy the wisdom of the wise, and the cleverness of the clever I will thwart." . . . For Jews demand signs and Greeks seek wisdom, but we preach Christ crucified, a stumbling block to Jews and folly to Gentiles, but to those who are called, both Jews and Greeks, Christ the power of God and the wisdom of God (1:18–19, 22–24).

Paul was teaching that those who are perishing—who experience their lives as heading toward death—do not recognize that the lesson of the cross saves us from despairing about our mortality. Christian faith, which ultimately does not rely on signs or logic, encourages us to respond to the mystery of human existence with faith that God has put us here for important reasons, such as to serve God and creation. Such God-directed living gives us a sense of inner purpose and makes us feel alive. In contrast, if we focus our lives on satisfying one acquisitive mimetic desire after another, we are more inclined to *experience* death on a daily basis. If our lives lack direction or meaning, we will desperately strive to feel alive by seeking pleasurable bodily experiences. Consequently we will *experience* the decay of our bodies with anger, disgust, fear, and loathing.[15]

In Fyodor Dostoyevsky's novel *The Brothers Karamazov*, Father Zossima teaches that loving and serving God's creation is a path to contentment and joy, as well as an appropriate response to the awe and wonder that accompanies experiencing the world's mysteries:

> Love all God's creation, the whole and every grain of sand in it.
> Love every leaf, every ray of God's light. Love the animals,
> love the plants, love everything. If you love everything, you will
> perceive the divine mystery of things. Once you perceive it, you
> will begin to comprehend it better every day. And you will come
> at last to love the whole world with all-embracing love. Love
> the animals: God has given them the rudiments of thought and
> joy untroubled. Do not trouble them, don't harass them, don't
> deprive them of their happiness, don't work against God's
> intent. Man, do not pride yourself on superiority to the animals;
> they are without sin, and you, with your greatness, defile the
> earth by your appearance on it, and leave the traces of your
> foulness after you—alas, it is true of almost every one of us!"[16]

Father Zossima's exhortations suggest insights into the existential questions that invariably challenge the self-conscious human mind. Why do we exist at a certain place and at a certain time? What is it about our own, individual identity that makes it ours and not anybody else's? Where did we come from? The scientific method cannot answer such existential questions. Science can describe associations between brain matter and brain functions, but such information does not adequately explain, for example, my own subjective experiences. A reasonable hypothesis for my own subjective, conscious existence is that I was created by some kind of divine power that I call "God." I believe that God cares about my existence, and if that is the case, it is reasonable to believe that I am here to serve God by making God's creation a better place. This faith accords with my experiences of finding constructive, compassionate, loving activities and relationships meaningful and enjoyable. In *The Brothers Karamazov*, Father Zossima gained similar faith through his experiences and through prayerful reflection on the life and teachings of Jesus as described in the Bible. I think that to the degree that we internalize a faith in God's love, we relieve our fear of death, neutralize our desire to build self-esteem at the expense of others, and experience life as a miraculous gift.

The critical component of my faith is not that God exists, because God's existence alone would not guide my life. Rather, the core of my faith relates to what I believe is God's nature. I believe God's essence is love and, I am convinced, this was also the faith of Christ. Such a faith encourages us to respond to satanic desires and satanic activities with love and compassion. This can be very difficult, because we often find ourselves physically and mentally wounded by life in general and hurtful people in particular, and we fear further wounds. Therefore, it is much easier to articulate this faith than to consistently act according to it. Our wounds often create barriers to our expressing God's love. Jesus taught us how to receive healing and to heal others.

Chapter 9: Healing

Healing in the Synagogue

I think one reason the Bible relates stories of Jesus healing so many people is that our broken world desperately needs healing. Dramatizing the urgent need for healing, Jesus often made a public spectacle of healing people with chronic (Luke 6:1–6; 13:10–16) or nonemergent maladies on the Sabbath (Luke 14:1–6). What has been the illness that has always stricken humanity? No doubt, people have long suffered from diseases such as infections, cancers, and heart disease. But from the standpoint of human community, the leading disease has been violence. Humans have killed untold millions of each other—approximately 160 million in wars during the last century alone.[1] Human violence has also blighted God's nonhuman creation, and every year humans unnecessarily abuse and kill many billions of animals.

Jesus taught his followers how to heal both the body and the soul, and such holistic healing is essential in healing broken relationships in our communities. In Mark's Gospel, Jesus began his ministry by entering the synagogue and healing a man with "an unclean spirit" (Mark 1:23–25). There are several remarkable aspects to this story. First, Jesus healed the man's demonic possession without harming the man. Often, humans had eradicated "demonic possession" by murder or banishment, encouraging people to hide any evidence that they might be possessed themselves. Many of those who find themselves irresistibly drawn to sex, alcohol, drugs, power, gambling, or fame experience their addictions as demonic possession—feeling as if forces greater than themselves compel them to behave in self-destructive ways. However, to maintain self-esteem and positive regard from peers, people tend to avoid acknowledging their addictions to themselves or others. Because of these personal and cultural barriers to healing, many of us need healing facilitators, such as a nonjudgmental, nonpunitive therapist or pastor.

One reason that it is difficult for therapists and pastors to heal is that addictions almost always have roots in past interpersonal relationships and experiences with family members and other members of society. Sometimes, the therapist or pastor can alleviate the symptoms of demonic possession to the point that the person can function in society at large. Completely healing the individual requires members of the larger community to face their own demons, which may have contributed to the individual's demonic possession. However, community members have tended to deny that they are plagued by demons of their own, and instead they have tended to project their own shameful passions onto unfortunate scapegoats.

Conveniently, some people assume the status of a scapegoat. These people are often deeply wounded spiritually or psychologically, and, feeling overwhelmed by their demons, they are unable to hide the manifestations of their possession from public view. The community perceives only these unfortunate few as possessed, even though everyone, more or less, is possessed by demonic passions. The difference is that most of us can repress or project our demons, and, consequently, most of us appear "healthy" and "normal" to the outside world. Jesus was able to heal the possessed man, rather than scapegoat him, because Jesus had already faced and rejected the three temptations that would have encouraged Jesus to project his own internal demons onto the scapegoat.

A second remarkable aspect to Jesus' healing the possessed man is that this healing occurred in the synagogue, where Jesus repeatedly cast out demons (Mark 1:39). Only clean people were allowed in the synagogue, so people with "an unclean spirit" would normally be excluded. Historically, religions have distinguished "clean" from "unclean" people, because people are eager to convince others (and themselves) that they are clean and justified in the eyes of God. Such delusions of purity often make people feel entitled to condemn others as possessed by "unclean spirits," "evil inclinations," or "sinful desires." It is not clear how someone identified as having an unclean spirit got into the synagogue, but the important point was that Jesus did not do what was customary at the time and expel the man. Instead, he chose to expel the demon.

A third point relates to how those in the synagogue received Jesus. Before Jesus healed the possessed man, they admired his teaching: "And they were astonished at his teaching, for he taught them as one who had authority, and not as the scribes" (Mark 1:22). Those with authority had been able to "cure" the problem of demonic possession only by killing or expelling the possessed person. However, Jesus healed with authority in a new way: "And they were all amazed, so that they questioned among themselves, saying, 'What is this? A new teaching!'" (1:27). Remarkably, their amazement was not focused on the healing per se. Rather, they were astonished by the *teaching*, saying, "With authority he commands even the unclean spirits, and they obey him" (1:27). The authorities had cured possession by scapegoating; Jesus' nonviolent solution to the problem of demonic possession was a new teaching.

Holistic Healing: The Man with Leprosy

Scapegoating invariably involves having "insiders" and "outsiders." Jesus challenged the legitimacy of these distinctions by healing in the synagogues, where only "clean" people were allowed, and by going so far as to touch a man with leprosy (Mark 1:40–45).

The ancient Hebrews believed that disease reflected God's judgment; consequently, they saw leprosy as a sign of sin. The man with leprosy was rejected by his community, and Jesus, "moved with pity . . . stretched out his hand and touched him" and made him clean. Jesus told the man to go directly to the priest "and offer for your cleansing what Moses commanded, for a proof to the people" (Mark 1:44).[2]

In ancient Hebrew culture, as in other primal cultures, touching an unclean person both rendered an individual unclean and made the individual an outsider. Thus, the people believed that Jesus became an outsider when he touched and healed the leper. Jesus had told the man with leprosy that, having been cleaned, he should "say nothing to any one." But "he went out and began to talk freely about it, and to spread the news, so that Jesus could no longer openly enter a town, but was out in the country; and people came to him from every quarter" (Mark 1:45). Because many people now regarded Jesus as unclean, Jesus was forced to reside in the countryside, evidently temporarily, because he later visited cities. Those who recognized their need of healing, unlike the members of the crowd, still sought Jesus' ministrations.[3]

The instruction to offer at the Temple "what Moses commanded" evidently relates to the animal sacrifices involved in the ritualistic cleansing of people with leprosy described in Leviticus 14. Nonetheless, I do not think that this story shows Jesus' endorsement of animal sacrifice. As Jesus likely expected, evidently the man did not comply with Jesus' instruction to go to the Temple. The man, having been cleansed by Jesus, did not need cleansing in the Temple, and he had a motivation to forgo the Temple ritual. In the Temple, the cleansing ritual involved shaving the head and eyebrows, as well as performing animal sacrifices. The eyebrows grow back very slowly, and the man probably would not have wanted to be marked for years as a former leper. Therefore, by cleansing the man, Jesus made ritual cleansing unnecessary; by instructing the man to follow the law, Jesus avoided scandalizing the religious authorities. This explanation, although admittedly speculative, retains an image of Jesus as loving, merciful, compassionate, and wise.

Today, physicians rely heavily on the biomedical model that views disease in terms of dysfunction of one or more body parts. However, the biomedical model often fails to address the psychological, spiritual, and social aspects of illness. Jesus exemplified holistic healing, which includes eradicating shame and social isolation. Jesus reintroduced the man with leprosy into the community by several means: Jesus first touched the man, signaling his regard for the man's worth; Jesus then healed the man's visible lesions; finally, Jesus declared him clean, making ritual shaving unnecessary.

Healing and Empathy

The Bible relates that Jesus was, with God's help, able to raise Lazarus from death (John 11:41). Jesus wept upon visiting Lazarus' grave (John 11:35), and this illustrates how sentiment inspires action. It seems that most children have a natural empathy for humans and animals, recognizing when humans and animals are joyful or suffering. Callousness and cruelty require somehow repressing our natural empathy and hardening our hearts. How does this happen?

It is instructive to consider how people learn to victimize animals. Children generally like animals, and adults consider children's kindness to animals a virtue. However, acculturation in our society involves restricting affection and concern for animals, perhaps because nearly all adults participate directly or indirectly in animal suffering and death. Widespread concern for animal welfare could interfere with contemporary animal use in agriculture, clothing, experimentation, hunting, and entertainment. How do animal-loving children grow up into adults who acquiesce to or even endorse animal abuse?

Based on numerous conversations, it seems that frequently children, upon learning that hamburgers come from cows and that "chicken" is actually a part of a chicken, express a desire not to eat meat. In many households, parents sternly respond that the child must eat the meat or forgo dessert. When this happens, most children resolve the conflict between their heart and their stomach by training their mind not to equate the meat on their plate with animals. However, as Christians we must acknowledge that whenever we hide the truth, we open the path to sin: "For every one who does evil hates the light, and does not come to the light, lest his deeds should be exposed" (John 3:20).

Those who live with farmed animals face particular difficulties. It is easier to suppress mental images of animals when one only sees flesh under cellophane; children on farms must interact with and come to know the animals who will be killed and eaten. Programs like 4-H, a USDA program for youth, help to eradicate children's usual love for animals. Many 4-H participants take infant animals and raise them to "market" size. The children care for the animals, and often the children and the animals love and trust each other. For many children, their betrayal of a loving, trusting friend is a traumatic experience.[4] Subsequently, those children will likely be either wracked by guilt and self-loathing, or, more commonly, they will repress feelings of empathy for animals and come to see all farmed animals as "things" meant to be slaughtered and eaten.

Animal and human abuses typically involve using terms that deny victims their individuality in order to objectify or demonize them. Killers during the Rwandan genocide called their victims "cockroaches," and people often use animal names such as "pig," "chicken," and "cow" as epithets to express contempt.[5] Notably, the animal names people employ to express disregard are those animals whom humans eat or harm in other ways. Objectification and demonization typically accompany injustice, whether the victims are humans or animals.

Spiritual Healing: The Invalid Man

John 5 describes Jesus healing a man who had been an invalid for 38 years. The man was among "a multitude of invalids, blind, lame, paralyzed" who were at a pool "by the Sheep Gate" (John 5:2), which was the gate through which the sheep destined for sacrifice passed. Gil Bailey has asserted that the juxtaposition of these people and the Sheep Gate was not accidental.[6] Ascribing guilt to those with infirmities is a kind of scapegoating, because people regarded infirmity as a sign that the infirm person or an ancestor had sinned.

What I find most remarkable about this story is the scene in the Temple subsequent to Jesus' healing. Jesus found the former invalid there and said to him, "See, you are well! Sin no more, that nothing worse befall you" (John 5:14). The man evidently planned to make a sacrifice in the Temple, presumably to thank God for his good fortune and to reduce the risk that he would once again receive God's wrath in the form of an infirmity. However, Jesus said, "Sin no more." What was the man's sin? I do not think the man's sin was related to his previous infirmity. Jesus rejected sin as an explanation for infirmity when he declared that a man's blindness was unrelated to his own sin or that of his parents (John 9:3). Disease can be a consequence of sin (e.g., gluttony), but the notion of disease as divine punishment for sin seems to run counter to the image of God as loving and forgiving. Furthermore, if those with diseases and infirmities were suffering the consequences of divine punishment, then evidently Jesus' healing would be undermining God's will.

Given the central role of sacrifice in the Temple, I think it is reasonable to conclude that the sin to which Jesus referred was the sacrifice itself. The man's experience of God's power through Jesus should have provided faith in God's love and goodness. Sacrifice as a means to approach God or to thank God shows a lack of faith in God's unconditional love. Consequently, Jesus warned the man to cease sacrificing, lest something worse befall him. That something is having one's life grounded on the lie that we can curry God's favor by sacrificing innocent victims. Wholeness is much more about living with integrity than about being able to walk. Therefore, the story describes how, after this meeting with Jesus in the Temple, the man told everyone how Jesus had healed him (John 5:15). Formerly disabled, the man was now truly whole, both physically and spiritually.

Healing and Faith: The Woman with Perpetual Bleeding

The story of the woman with perpetual bleeding provides important insights into the nature of healing. The version of Luke reads,

> And a woman who had had a flow of blood for twelve years and could not be healed by any one, came up behind him, and touched the fringe of his garment; and immediately her flow of blood ceased. And Jesus said, "Who was it that touched me?" When all denied it, Peter said, "Master, the multitudes surround you and press upon you!" But Jesus said, "Some one touched me; for I perceive that power has gone forth from me." And when the woman saw that she was not hidden, she came trembling, and falling down before him declared in the presence of all the people why she had touched him, and how she had been immediately healed. And he said to her, "Daughter, your faith has made you well; go in peace" (Luke 8:43–48; see also Matthew 9:20–22; Mark 5:25–34).

According to Jewish law, menstruating women were unclean and could not touch anyone or be touched. When Jesus asked who touched him, the woman was afraid, because she feared Jesus would be angry at her violation of the law. However, Jesus did not express disgust or revulsion; rather, he said only that he sensed power going forth from him.

The woman was compelled to confess because she knew that Jesus could identify her. However, in addition to acknowledging her act, she declared that she was healed. This, I think, is what Jesus recognized as her faith. Though he had participated in her cure, he did not say that he had healed her. Instead, he observed that her faith, which had inspired her to publicly declare that Jesus had healed her, had made her well.

This story illustrates an important point about disease. Disease (dis-ease) is a state of mind in which one is discontented with some aspect of the body. One can have a dysfunctional body and not be diseased, and one's body may be functioning quite adequately yet a person may *experience* disease. Everyone has the spiritual need for a sense of direction and purpose in life, which often involves a sense of connection to God. If our spiritual needs are not met, then we are prone to suffer existential anxiety and, consequently, to feel diseased even when our bodies work well. The woman's faith made her well enough to align herself with Jesus, and she was prepared to "go in peace." Likewise, after Jesus healed ten people with leprosy, only one returned to thank him; Jesus said to him, "Your faith has made you well" (Luke 17:19). Jesus healed the man's disease, and the faith of Christ healed his mind and spirit.

Healing the Gerasene Demoniac

Mark 5:1–20 and Luke 8:26–39 (see also Matthew 8:28–34) relate the story of the Gerasene demoniac. Jesus exorcised demons from a possessed man, and the demons then inhabited a herd of swine who ran crazed down a steep bank and drowned (Matthew 8:32; Mark 5:13; Luke 8:33). Girard has argued that the Gerasene demoniac story reveals profound insights into scapegoating.[7]

The possessed man was the communal scapegoat. He bore the burden of the people's unclean spirits: they could blame him for their own forbidden thoughts and desires that threatened social order and peace. In Mark's account, "Night and day among the tombs and on the mountains he was always crying out and bruising himself with stones." Normally, those deemed to be possessed were hunted, stoned, and killed. Here, the man hid in the tombs and stoned himself, protecting himself from the scapegoat's usual fate. His howling was an affront to them, but they did not kill him. Rather than kill him, they bound him in chains. These chains were insufficient to hold him, which allowed him to bruise himself with stones without causing lethal damage. His self-injury satisfied the community's need for a scapegoat.

And so, there was a balance between the insufficient chaining by the community and the insufficient self-stoning by the man. This balance allowed the possessed man to live while serving the community's need for a scapegoat. Perhaps this unusual arrangement began when the scapegoat, recognizing that angry communal members were convinced of his demonic possession and were determined to stone him, started to stone himself. Because it seemed that the "demons" were already stoning the man, the community refrained from joining the demons in stoning him.

The possessed man naturally feared Jesus, who had said, "Come out of the man, you unclean spirit!" (Mark 5:8). Jesus threatened the balance of violence between the man and the community, which could have led to the man's death. When Jesus asked their names, the demons replied, "My name is Legion; for we are many" (Mark 5:9). The demons represented all the forbidden desires of the community. They were parts of the human psyche, and they did not have individual names that would have indicated that they existed independently. The demons begged Jesus not to send them out of the country (Mark 5:10; "into the abyss" in Luke 8:31). I think that the request of the demons reflected the community's concern that exorcising the demons would have forced the community to find a new scapegoat onto whom they could project their guilt, fears, hatreds, and illicit desires.

Frequently, people have tried to transfer the role of the scapegoat from humans to animals. When the demons asked to be sent into the swine, this represented the community's desire to see its own demons find a new home. The story refers to swine as the recipients of the spirits, because swine, as unclean animals in Jewish eyes, seemed appropriate repositories of unclean spirits. However, the crazed swine went over a steep bank and drowned. This reverses normal human patterns of behavior. According to Girardian theory, typically people have metaphorically or literally thrown those they regarded as possessed off a cliff. However, in this story, the possessed man was saved; and the demons that had afflicted the community, and that had been projected onto the possessed man, were destroyed.

The community's response to the cured man is illuminating: "They were afraid" (Mark 5:15). Their scapegoat was cured, and, consequently, their peace and equanimity were threatened. Some commentators have argued that the people were upset about the economic loss of the pigs, but if that had been the case, the people would have been angry, not afraid. The Gerasene people asked Jesus to depart because he had damaged the social order. Meanwhile, the cured man begged to leave town with Jesus, most likely because the man was at high risk of being stoned by a community desperate to reestablish order. But Jesus refused the man's request, forcing the man to bear witness to Jesus' method of healing by destroying demons rather than by destroying people. People marveled at the cured man's story, indicating that destroying demons was not as socially devastating as everyone had feared.

I see demons as those desires that separate us from God and from each other. Because our desires are mimetic, our demonic desires do more than possess us individually—they can become enshrined in institutions. Persons inspired by the faith of Christ, whether they are Christian or not, work to cure the demonic possessions that afflict individuals and institutions. However, institutions can be more difficult to cure, because they can become false gods to which people offer blind allegiance. Consequently, one needs to demonstrate that institutional demons derive from and depend on lies. Jesus' self-sacrifice on the cross showed that "sacred" violence enshrined in religious "laws" is scandalous.

The analysis of the Gerasene demoniac story contains the troubling implication that Jesus was responsible for the death of 2,000 innocent pigs. An interesting aspect of this story is that there is no such "steep bank" (Matthew 8:32; Mark 5:13; Luke 8:33) or sea near Gerasa (Mark 5:1; Luke 8:26) or Gadara (Matthew 8:28).[9] Perhaps including this detail communicated to readers that one should regard this story as allegory rather than as literal historical narrative. Further evidence that the story is an allegory is that it is hard to imagine such a large herd as "about two thousand" pigs (Mark 5:13). Pigs, who do not have the strong herd instincts of sheep, would be much more inclined than sheep to wander off.

If one regards this story as an allegory, one can see how it uses metaphors that were familiar to the ancient Hebrews. The Hebrews regarded pigs as unclean; and the Hebrews would have understood the image of evil spirits going to the bottom of the sea, where many ancient people believed evil spirits reside. If taken as allegory, this story relates important aspects about Jesus' ministry while retaining a conviction, well-grounded in Scripture, that Jesus cared about all God's creation.

Healing a Broken World: The Man Born Blind

John 9 describes Jesus healing a man born blind.[9] I would like to highlight several remarkable features of this story that relate to how Jesus' ministry was fundamentally a healing ministry.

The text reads, "And his disciples asked him, 'Rabbi, who sinned, this man or his parents, that he was born blind?' Jesus answered, 'It was not that this man sinned, or his parents, but that the works of God might be made manifest in him' " (John 9:2–3). Jesus rejected two widespread beliefs. Although Exodus 34:7 had declared that children are punished for the sins of their parents, Jesus taught along the lines of Ezekiel 18:14–17, that God does not visit punishment on the children of sinners. Furthermore, Jesus disputed the commonplace notion that disease is a sign of sin. Paul observed that all of us fall short of the glory of God (Romans 3:23), and if God were wrathful and punished sinners, there would be no good reason to spare any of us.

Jesus then said that he was doing the works of God, indicating that creation is not complete. This notion was reinforced by Jesus healing the man on the Sabbath, which angered the Jewish authorities, even though there was no urgency to heal the man. Similarly, after Jesus healed a paralyzed man he said, "My Father is working still, and I am working" (John 5:17). Interestingly, Jesus healed the blind man with dirt, which harks back to Genesis 2:7, in which God created man with "dust from the ground." Jesus participated in God's work of completing creation.

Completing creation involves reconciling the world to God's original intentions, a world in which all creation lives peacefully and harmoniously (Genesis 1:29–30; see also Isaiah 11:6–9). In order to reconcile creation, Jesus would need to "take away the sin of the world." Informed by Girardian mimetic theory, I have suggested that the "sin of the world" is scapegoating. Our world will always be broken, violent, and in need of healing as long as our communities are grounded on and maintained by scapegoating. Christianity teaches us that love and forgiveness are the proper foundations of a community of peace and harmony.

Christians are called to help heal a broken world, and, by doing so, join Jesus in reconciling creation (2 Corinthians 5:19). Healing involves restoring spiritual, as well as physical, wholeness. Spiritual wholeness requires acceptance into community, partly because we are social creatures; partly because, in order to serve God, we need to serve others; and partly because our participation in and acceptance by community reminds us that we are all God's beloved children. Therefore, Paul wrote, "There is neither Jew nor Greek, there is neither slave nor free, there is neither male nor female; for you are all one in Christ Jesus" (Galatians 3:28).

The universal fear of death is often a major stumbling block to our participation in the reconciliation of creation. Jesus recognized this when he said, "For whoever would save his life will lose it, and whoever loses his life for my sake will find it" (Matthew 16:25; see also Matthew 10:39; Mark 8:35; Luke 9:24, 17:33). The reason, I think, relates to the quest for self-esteem as a salve against the universal fear of death. If we do not ground our self-esteem in our relationship to God, we can only gain self-esteem by being superior in relation to other individuals. In practice, being superior often involves victimizing vulnerable individuals in an attempt to gain power, wealth, or whatever one's culture regards as valuable. However, no amount of self-esteem can fully eradicate the fear of death. Even though humans can repress their fear of death from consciousness, death's inevitability haunts the subconscious mind. Consequently, the typical human response to mortality fears has been to compulsively, relentlessly seek *more* self-esteem. Never having enough self-esteem to quell death anxieties, even those who "should" be happy with their degree of "success" tend to find themselves perennially dissatisfied with their lives. Therefore, our human desire to save our life (i.e., to gain enough self-esteem to overcome the fear of death) causes us to fall into conflict with and become disconnected from God's creation, which in turns alienates us from God. The desperate attempt to save our life distances us from God, the source of life, and increases our sense of mortality. And so, the project to save our life results in our losing it.

Jesus taught that we should trust in God's love and goodness and surrender our life to God. This was Jesus' cure for the "dis-ease" engendered by the fear of death that troubles the human mind, disrupts communities, and leads to scapegoating.

The stories about the life, death, and resurrection of Jesus demonstrate that we do not need to fear death. If we believe in a loving God, it follows that whatever happens to us when our physical body dies, we should not expect death to be bad. If fear of death does not rule our lives, we can more courageously face the dangers that can accompany being healers of physical and social maladies in a broken and violent world.

Healing: A Christian Calling

Healing is one way to answer our calling to express love. But what can we do, specifically, to help heal? We can listen, which shows that we care; we can respectfully offer what help we can; and we can help heal with respectful, appropriate touch, such as when Simon Wiesenthal let a dying, confessing Nazi hold his hand and confess his sin. Often, the greatest healing occurs when we help people understand that they matter to God. Although only God can fully heal the soul, we are called to help, and our life experiences provide valuable tools.

To widely varying degrees, we have all been wounded by life. We have all experienced loss, and we have all experienced the crushing feelings associated with humiliation. We know what it feels like to be wounded, whether intentionally or unintentionally, by family, friends, strangers, and bad luck. And we have tried to develop strategies to deal with painful memories and to prevent further wounds. Often, the most deeply wounded people have rejected God, either because they have internalized their status as scapegoats and believe they are unworthy of God's love, or because they have found it impossible to believe in a loving God who has failed to protect them. Our experiences help us empathize with other wounded people, even if their pain is far deeper than what we have experienced. Our empathy makes it possible for us to connect with wounded people intellectually, emotionally, and spiritually, which in turn helps us heal other wounded individuals holistically.

Healing almost always is a communal activity. We are social creatures, and throughout our lives our social interactions shape and modify our sense of identity—who we are and how we relate to the larger universe. Our relationships profoundly affect whether we have a sense of meaning, whether we have good self-esteem, and whether we feel connected to or alienated from the world. Our sense of personal health is strongly linked to the health or pathology of our relationships. We can have physical infirmities and still feel valuable and whole, and we can enjoy good physical health and still suffer from a sense of alienation and despair. Therefore, healing the body, mind, and soul is a communal activity, which reinforces the need for spiritual communities dedicated to mutual care, support, and healing. For Christians, church communities are the vehicles through which we can give and receive the fruits of the faith of Christ.

How can we help heal those individuals who are unable to participate in communities of faith, such as people with mental disabilities or animals? Sometimes, we can help heal through mere presence or touch. Sometimes, we can help heal from afar, by mobilizing efforts to change the conditions that wound them. Many people believe that prayer can help heal by directing God's healing energy toward those in need.

Whatever we do to reconcile God's creation to the biblical ideal of peaceful, harmonious co-existence is a healing ministry. And healing often involves being a peacemaker.

Chapter 10: Peacemaking

What is Violence?

For purposes of this discussion, I will take *violence* to mean intentional, unnecessary harm by individuals or institutions. Violence by individuals includes physical or emotional harm. I can envision situations in which physical force is not violent, such as animal aggressiveness that emanates from biological or instinctive drives. Likewise, one might also regard as not violent those activities that cause physical harm but are necessary to preserve an individual's life, such as defending oneself or others against attack, or hunting when there is no alternative sources of nourishment available.

Unlike violence by individuals, institutional violence generally does not require ill intent by those perpetrating the violence. Institutions are often grounded on the scapegoating process, and the original perpetrators might or might not have been aware of the violence and injustice they were inciting. Those who continue to perpetrate institutional violence often believe their activities are righteous and just. They are abiding by the institution's myths that hide both the original violence and the ongoing injustice. Many of those enforcing "Jim Crow" segregation laws were taught from childhood that blacks were inferior to whites and that segregation benefited both blacks and whites.

It is often difficult to distinguish between legitimate use of force and illegitimate violence, because people generally regard their own violence as necessary for "justice," "self-defense," or to preserve "sacred" institutions. Those who genuinely regret any harm they have caused, and have done their best to limit harm, probably have used force out of necessity. We can be more confident that our use of force is not violent if we are trying to protect other individuals, rather than promoting our own interests, restoring the "dignity" of our community, or avenging perceived wrongdoing. Those who take pride in their triumphs over what they call "evil," or grab at the spoils of victory, have likely engaged in acts of violence.

If our *intent* is loving and compassionate, we will tend to avoid participating in violence. Proverbs relates, "Every way of a man is right in his own eyes, but the Lord weighs the heart" (21:2), and 1 Samuel 16:7 reads, "Man looks on the outward appearance, but the Lord looks on the heart." Similarly, Jesus taught, "Blessed are the pure in heart, for they shall see God" (Matthew 5:8), and Jesus often emphasized the importance of intent (Matthew 5:28, 6:1–6, 18:35; Mark 12:42–44).

We should always be uncomfortable with activities that harm other individuals. We should question our own motives repeatedly, and we should constantly seek to view situations from victims' perspectives. If we call our harmful activities revenge, purification, or divine sacrifice, then it is likely that we have obscured our violence behind mythological stories that attribute our violence to God or to a secular ideology, such as nationalism. S. Mark Heim has written, "To veil it [violence] under euphemism and mythology, to be piously silent before its sacred power, is to make its rule absolute."[1]

Was Jesus ever violent? The only biblical story in which Jesus used physical force against adversaries was in the Temple when he turned over the tables of the money-changers (Matthew 21:12–13; Mark 11:15–17; John 2:14–16). Importantly, though Jesus' words and actions suggest anger, Jesus did not hurt anybody. Why did he disrupt their activities? Some believe that the money-changers were cheating the pilgrims who needed to change currency to buy animals for sacrifices. It is possible that some money-changers cheated unsuspecting pilgrims, but it seems unreasonable to believe that Jesus would anger the Roman authorities and the powerful Temple priests only to prevent petty crimes.

I think Jesus objected to the sacrificial system, as did many antisacrificial latter Hebrew prophets. Remarkably, in John's account, Jesus also drove out the animals slated for sacrifice. I do not regard Jesus as acting violently. No individuals were injured, and Jesus' liberation of the animals was necessary to prevent grave harm to innocent individuals.

"I Desire Mercy and Not Sacrifice"

The Bible's many references to sacrifices challenge the Girardian notion that the Bible aims to teach people how to build community without scapegoating. The Hebrew Scriptures seem to provide conflicting views on sacrifice. The earlier writings described human sacrifice and instructions from God concerning animal sacrifices. Several latter prophets condemned all blood sacrifices, and Jesus recalled Hosea 6:6 when he said, "Go and learn what this means 'I desire mercy, and not sacrifice.' For I came not to call the righteous, but sinners" (Matthew 9:13). In this passage, Jesus defended his eating with tax collectors and sinners whom the people scapegoated by ostracizing them.

Similarly, in Matthew 12:5–7, Jesus responded to the priests who had criticized his disciples for plucking heads of grain on the Sabbath to eat:

> Have you not read in the law how on the sabbath the priests in
> the temple profane the sabbath, and are guiltless? I tell you,
> something greater than the temple is here. And if you had known
> what this means, 'I desire mercy, and not sacrifice,' you would
> not have condemned the guiltless.

Though Jesus was specifically referring to his guiltless disciples, I think that Jesus' comment was meant to apply to all innocent victims condemned by the priests, including the totally guiltless sacrificial animals. These animals were the concern of Hosea 6:6, the passage upon which Matthew 9:13 and 12:7 is based. Further evidence that Jesus included animals among those who were condemned but guiltless was his comment "something greater than the temple is here" (Matthew 12:6)—animal sacrifices were a central function of the Temple.[2]

Some have claimed that Jesus did not object to sacrifices per se but rather to those who performed sacrifices while remaining hard of heart and sinful. If this were so, it would have made more sense for Jesus to say, "I desire mercy more than sacrifice" rather than "I desire mercy and not sacrifice." Evidence that Jesus objected to all sacrificial violence comes from an exchange between Jesus and a scribe about what is the first commandment. The scribe said,

> To love him with all the heart, and with all the understanding,
> and with all the strength, and to love one's neighbor as oneself,
> is much more than all whole burnt offerings and sacrifices
> (Mark 12:33).

The text continues, "And when Jesus saw that he answered wisely, he said to him, 'You are not far from the kingdom of God' " (Mark 12:34). Perhaps Jesus would have concluded that the scribe had reached the kingdom of God if the scribe had *fully* rejected "whole burnt offerings and sacrifices" rather than only saying it is much more important that we love fully.

This analysis offers an interpretation of Romans 12:1, in which Paul writes, "I appeal to you therefore, brethren, by the mercies of God, to present your bodies as a living sacrifice, holy and acceptable to God, which is your spiritual worship." The age of sacrifice had ended, and Paul wrote that we are to dedicate ourselves completely, including our bodies, to God. This passage, I think, helps us better understand Romans 6:23: "For the wages of sin is death, but the free gift of God is eternal life in Christ Jesus our Lord." Many people have interpreted Romans 6:23 as indicating that, as a consequence of sin, God demands the death either of the sinner or of a sacrificial substitute—and the ultimate sacrifice was Jesus. However, because Romans 12:1 points to *self*-sacrifice, I do not think we should read Romans 6:23 as an indication that God desires that we sacrifice other individuals to substitute for ourselves. Indeed, Romans 6:23 does not say that God desires death at all. The passage has made a simple and valid observation: sinfulness leads to death. If we sin by failing to express God's love, we fall into rivalries that lead to violence and death.

Violence and the Churches

The Hebrew Scriptures describe God's ideal as peaceful, harmonious coexistence throughout God's creation. There was no violence in the Garden of Eden, and Isaiah 11:6–9 prophesied a return to this harmonious state. Isaiah foresaw a time in which "they shall beat their swords into plowshares, and their spears into pruning hooks; nation shall not lift up sword against nation, neither shall they learn war any more" (2:4). Similarly, Jesus encouraged nonviolence (Matthew 26:52).

As Christianity evolved from a movement to reform Judaism into a distinct religion, it developed a hierarchical establishment that has sometimes lost sight of Jesus' ministry. Those with power have been tempted to defend their own privileged positions and to promote their personal agendas, rather than to dedicate themselves to serve God by helping to heal a broken world. Churches have important religious and social functions, but there is always the danger that churches, like all institutions, can participate in victimization and scapegoating.

Girard has maintained that all hierarchies have their foundations in the scapegoating process. If true, it follows that hierarchies are grounded in violence and maintained by violence, though the violent elements may be subtle or hidden. In churches and other hierarchical institutions, those with less power usually abide by the dictates of those with more power. Often those with less power yield out of respect for the "sacred" power arrangements, and defying those with power is taboo. At other times, those with less power acquiesce, because the forces supporting the hierarchy are too strong. These forces ultimately rely on the threat of violence, such as from police who uphold laws that defend the power arrangements, or from a mob that would be scandalized by disrespect for the "sacred" order. Most conflicts between leaders of hierarchical institutions and subordinates are settled without resorting to violence, but physical force is the final arbitrator of all disputes in hierarchical social arrangements.

Although hierarchical institutions can help communities maintain order, and well-governed institutions can make intelligent decisions, there is always the risk that those with power will abuse their positions and serve their own ends rather than those of the community. How can Christians assess whether or not their church leaders are serving God and their communities? One way is to remain mindful that church authorities who seem to be promoting violence and destructiveness might not be preaching the Christian gospel. One should be skeptical of church authorities who declare that God's wrath and vengeance justifies violence and oppression "in the name of God." All too often, it seems, this is a strategy to intimidate or expel those who challenge church doctrines or power arrangements. Church authorities have often labeled as "enemies of God" those who have actually been enemies of the ambitions of church authorities.

The writer of 2 Timothy predicted that false teachers will attract people with stories that misrepresent God's nature and intent: "For the time is coming when people will not endure sound teaching, but having itching ears they will accumulate for themselves teachers to suit their own likings, and will turn away from listening to the truth and wander into myths" (4:3–4). Jeremiah similarly warned, "The prophets prophesy falsely, and the priests rule at their direction; many people love to have it so, but what will you do when the end comes?" (5:31).

Church-initiated violence can also occur when different religious bodies, including different churches, each claim to have the "one true faith." This readily leads to mimetic rivalries between churches than can divide communities generally and the body of Christ in particular. Bitter disputes between and within denominations undermine community building, when in truth diversity of theology and liturgy within Christendom can promote intellectual and spiritual growth. Christians should seek to become a unified body, bound together by a common goal to express God's love; and they should tolerate, or even celebrate, differences in theology or liturgy. Furthermore, if Christians are to promote peace and justice in our pluralistic society, they must respect people of other faiths. I have no quarrel with any persons committed to compassion and justice, regardless of the images they use to describe God, the spiritual leaders they revere, or the rituals that give them a sense of connection to the divine and each other. Such people have adopted the faith *of* Christ, even if they do not regard themselves as Christian and do not have faith *in* Christ.

Despite Jesus' nonsacrificial teachings, many Christian communities have yielded to the temptation to use scapegoating as the glue that holds them together. Some churches scapegoat people, such as homosexuals or religious skeptics, by claiming, I think inaccurately, that they threaten the church community. Similarly, there seems to be an element of scapegoating in many churches' attitudes about and treatment of animals. Many churches have emphasized humanity's importance by contrasting humans with animals. I think this is one reason that, in general, the churches have not been animal-friendly. Christian animal advocates have found that churches generally resist Christian education programs that aim to expose the massive suffering of billions of animals annually on factory farms; many churches celebrate killing animals with social events such as pig-roasts and fish-fries; and some churches even sponsor "Christian" hunting clubs.

Christians are called to witness for God as Jesus did, and this includes defending victims, human and animal, from violence. In a world of violence, Christians should strive to be peacemakers.

Meekness and Peacemaking

Jesus likely surprised those who listened to his Sermon on the Mount when he declared, "The meek shall inherit the earth" (Matthew 5:5). Everyday experience did not lend credence to this prediction. Indeed, back then—as today—the advice "If any one strikes you on the right cheek, turn to him the other also" (Matthew 5:39) likely seemed foolish, because it seemed to only invite more abuse.[3]

Meek humans have often been abused, and every year billions of meek animals experience grievous mistreatment by humans. Therefore, turning the other cheek seems a poor strategy for self-preservation. However, the Bible's eschatological, or end-of-time, vision—known as the "Peaceable Kingdom" or the "realm of God"—anticipates all creation living in harmony. Christian faith teaches us that love, compassion, mercy, and humility will eventually prevail, and the children of God will herald a new creation in which everyone will be free from bondage (Romans 8:18–25).

In hope and expectation of the realm of God, our decision to "turn the other cheek" is an act of obedience and faith. It might prove effective, because turning the other cheek is an act of nonviolent resistance that shames the abuser and may thereby bring about a transformation.[4] Ultimately, however, we do not know how "the meek" can prevail, and perhaps it will require divine intervention. For all creation to be in harmony, the wolf will live peacefully with the lamb and the lion will eat straw (Isaiah 11:7, 65:25). ✓

Is Peacemaking Practical?

Jesus said, "Behold, I send you out as sheep in the midst of wolves; so be wise as serpents and innocent as doves" (Matthew 10:16). Jesus, while using nonviolent means, skillfully avoided physical danger from mobs and deftly handled the theological traps set by the Pharisees. Dr. Martin Luther King, Jr. similarly exhibited determination tempered by prudence. He strictly observed nonviolence, which he considered essential on moral and practical grounds, even after racists bombed his home. He did not seek to become a martyr, but he understood the grave risks of his ministry.

In a world filled with wolves, it is tempting to be aggressive and violent. Though we must be prudent, we are called to remain innocent. In this way, we may be beacons of peace and love. Jesus taught that being innocent (i.e., harmless, loving, and nonjudgmental) can be dangerous, and I suggest at least two possible reasons. Aggressive people may interpret nonviolence as a sign of fear and weakness, which can encourage further aggression. Alternatively, they may be inwardly ashamed of their violence, and a nonviolent response heightens their discomfort. This makes them angry, and they project their anger onto the nonviolent people. They may accuse the nonviolent people of self-righteousness, hypocrisy, or even satanic possession, because violent people often believe that God endorses their own "righteous" violence.

Although Jesus suggested that peacemakers will ultimately prevail, historically they have often been victims of violence. From the martyred first century Christian pacifists[5] to the many victimized Quakers, peacemaking people have been harassed and even killed. Often, peacemakers have incurred wrath because, by refusing to join scapegoating mobs, they have threatened to expose the scandal of the scapegoating process.

Proponents of pacifism have pointed to successful peaceful movements, such as those led by Dr. King and Mohandas Gandhi. However, both were assassinated, and many of their followers were victims of violence. Furthermore, it could be argued that their success owed much to the fact that their oppressors feared violence from other quarters. Many whites preferred accommodating reforms advocated by Dr. King's nonviolent movement to the radical changes demanded by violent black leaders. Similarly, British forces occupying India feared rising nationalism among hundreds of millions of Indians. This encouraged a peaceful transfer of power rather than a bloody and costly war. The moral strength of Martin Luther King and Mohandas Gandhi encouraged effective nonviolent resistance, but the threat of violence from less peaceably minded people likely played an important role in their successes.

I think Jesus said "Blessed are the peacemakers," not because they will prevail in a physical sense—though sometimes they do[6]—but because they prevail in a spiritual sense. Christianity is not only about practical outcomes, as illustrated by Jesus' comment to Pilate, "My kingship is not of this world; if my kingship were of this world, my servants would fight, that I might not be handed over to the Jews" (John 18:36). Jesus was primarily concerned with serving God.

It is relatively easy to be nonviolent in the United States, where a powerful military protects its citizens from harm, and in other nations where there are no imminent threats of foreign invasion or internal revolt. Commitment to nonviolence is much more problematic, both practically and morally, in war-torn parts of the world, particularly if one has children to protect.

It is important to distinguish between being peaceful and being a peacemaker. Rulers have no quarrel with those who peacefully acquiesce, and peaceful people may avoid violence and destruction. Peaceful people can be ruthlessly exploited, but rulers will find no need to violently repress them. In contrast, peacemakers challenge the rulers and other temporal powers, and they often become victims of violence. Peacemakers know that violence underlies all unjust social arrangements, because maintaining injustice requires violence against those who demand justice. Peacemakers recognize that the only way to end violence is to reveal injustice and violence, which can be dangerous work.

Peacemaking and Christian Community

Luke 9:51–55 relates a story that illustrates Jesus' commitment to nonviolence. Jesus and his disciples were not welcomed in a Samaritan village "because his face was set toward Jerusalem." There were long-standing hostile feelings between Jews and Samaritans, and Jesus' disciples James and John asked Jesus, "Lord, do you want us to bid fire come down from heaven and consume them?" Jesus "turned and rebuked them." Jesus' ministry involved reconciliation, not retributive violence.

The biblical accounts of Jesus' life, death, and resurrection have shown that overcoming violence does not entail vanquishing enemies. Rev. Nuechterlein has written, "God's cure for violence is completely different than ours. God submits to our sacred violence in the cross and reveals it as meaningless and powerless compared to God's power of life."[7] Victory over violence and death involves participating in the reconciliation of creation, which is life-affirming and love-affirming.

Following Jesus helps individual Christians find peace in their hearts, and peacemaking also has a communal element, because violence is a communal problem with communal origins. It is important, but insufficient, for an individual to resolve to be a peacemaker. People must work together to find ways to redirect their mimetic desires from acquisitive mimetic desires that generate rivalries and conflicts toward desires that engender peace. Consequently, I see collaborating with members of one's community as a critical component of Christian living.

Jesus said that the way we live should be the means by which we spread the gospel: "Let your light so shine before men, that they may see your good works and give glory to your Father who is in heaven" (Matthew 5:16). We all have the potential to be "a light to the nations" (Isaiah 42:6, 49:6), which is essentially a prophetic calling.

Chapter II: Christian Faith and Prophetic Witness

Faith in a Living God

Christians understand faith in different ways, but I think a common denominator is that Christian faith involves regarding oneself as a child of God, because God is the source of all life.[1] Regarding ourselves as children of God encourages works that reflect reverence for God, the loving parent. We should avoid harming any of God's other creations—people, animals, and the earth.

Faith involves action, and James said, "So faith by itself, if it has no works, is dead" (2:17). Faith is only valuable and meaningful to God's creation if it inspires people to do good works. Indeed, James further explained, "I by my works will show you my faith" (2:18), and likewise Jesus said, "You will know them by their fruits" (Matthew 7:16, 7:20). Our relationships, our commitments, and our lifestyles, not our words or proclamations of faith, demonstrate what we believe. This, I think, is why Jesus said, "Not every one who says to me, 'Lord, Lord,' shall enter the kingdom of heaven, but he who does the will of my Father who is in heaven" (Matthew 7:21).

Faith in a living, active, caring God undermines the notion that there are no solutions to our growing social and environmental problems. One can believe that human spiritual and moral progress is possible, perhaps with the aid of the Holy Spirit. Jesus repeatedly showed great interest in his community's outcasts, such as women, people with disabilities, and tax collectors. His welcoming attitude was perceived as scandalous, and Jesus remarked— after describing his healing of blind, lame, leprous, deaf, and dead people, as well as his preaching to poor people—"Blessed is he who takes no offense at me" (Matthew 11:6).

The opposite of taking offense is having faith that one's way is right and true. Therefore, those with faith accept the possibility of receiving condemnation. This, I think, was the message encapsulated in the story of the Canaanite woman (Matthew 15:21–28). As a Canaanite, she was despised by many Jews, and after she asked Jesus to heal her daughter, Jesus compared her to a dog. She persisted in her request, showing that she was not offended, and Jesus, impressed by her faith, healed her daughter.[2]

Faith does not require certainty about tenets that, to the skeptical mind, seem very dubious. Faith involves trusting in God's goodness, which encourages us to live as if we can assist God in healing and reconciling creation. However, the existence of evil in the world challenges those who assert God's goodness.

God and the Existence of Evil

As discussed in Chapter 2, there should be no injustice if God is both all-powerful and righteous. Yet there appears to be widespread suffering and injustice, indicating that God is either not all powerful or not righteous. Let us briefly consider how some theologians have addressed this paradox.

Some deny that the world is unjust. Even though there is suffering, they maintain that this is nonetheless the best of all possible worlds. They assert that the reason it often seems that suffering and death are unnecessary is that we have a very limited view of God's greater design. If we more fully understood God's plan, we would recognize that everything is for the good. It is impossible to prove or disprove this theory, but I do not think it is reasonable. There is so much human and animal suffering in the world, much of it seemingly meaningless, that it is hard to believe that a righteous, all-powerful God wants it that way.

Another response is that God's notion of "the good" is very different from our own. However, if we believe that we should act according to God's will yet only have a vague and often mistaken notion of what God regards as good, then we are ill equipped to make sound moral decisions.

Some have maintained that God is not necessarily righteous. There is no reason, they assert, to assume that God had benevolent reasons for creating the universe and its living beings. Maybe the creator God derives pleasure from watching us struggle and suffer. Again, this is theoretically possible, but Rabbi Harold Kushner has said that this is not a god to whom he would pray.[3] One might perform rituals to appease such a malevolent deity, but one would not love and respect such a god. Kushner has held that God is not all powerful. Perhaps when a plane crashes, God is unable to save the kind and decent people who perish along with hateful people.

Some people have noted that if God were to directly intervene in human affairs and violate the physical laws of nature, this would deprive humans of free will. Our praising God and our acting according to God's will are meaningful only insofar as we have free will. However, I would question God's righteousness if God permitted such massive suffering in the world primarily because God desired to receive praise and dedication from humans. Even if human free will were necessary for human existence to be meaningful, I find this an insufficient reason to justify so much suffering of humans and animals. Further, humans and animals often suffer for reasons that have little, if anything, to do with human free will, including natural disasters, random diseases, and the commonplace suffering of animals in nature.

What if humans did not have free will and God controlled the workings of the world? In that case, I would doubt God's righteousness on the grounds that there is so much evidently unnecessary suffering. Indeed, any direct intervention by God into human affairs raises questions about God's righteousness. Let us say that God miraculously saved a child who was in an airplane that slammed into a mountain. God's saving this child would demonstrate that God has the power to spare people from otherwise certain death, and we would be forced to conclude that God generally chooses not to do so. Another example: After an airplane crash, the TV news sometimes features a shaken passenger who missed the plane. That person may conclude that God has spared him or her because of some special plan God has. Nearly every flight has at least one person who changed plans or missed the flight, and that is the "survivor" who ends up before the television cameras when the plane crashes. If God had really spared that person for a reason, then one must also conclude that God chose to allow the rest of the passengers to die for a reason. Many of those people were probably good, caring people who played important and valuable roles in others' lives, and God's allowing them to die would raise doubts about God's righteousness.

Why would God create such an imperfect universe? Though I do not know, by faith I believe that God cares about it. The alternative to this faith, it seems, is nihilism and despair. Perhaps, as Kushner has posited, God created a universe full of possibility that, once created, was beyond God's power to control. However, we do have the capacity to choose whether to side with victims or victimizers, and Christian faith indicates that God has the power to help guide us. How can we discern our calling to the "kingdom of God"?

Receiving the Kingdom of God "Like a Child"

Jesus said, "Whoever does not receive the kingdom of God like a child shall not enter it" (Mark 10:15; Luke 18:17). I think that one reason one must be "like a child" has to do with the nature of children's desires. As anyone who interacts with children knows, they are not totally innocent. They can be selfish and mean. However, children differ from adults in that children tend to be less complicated and less calculating about getting what they want, and they are less likely to carry a grudge if they do not get it. The last feature is important from the perspective of mimetic theory: Children, like adults, care about self-esteem, but in general their desires are more physical and less symbolic. A child seeing another child playing with a toy will often, because of mimetic desire, want to play with that same toy. Failure to obtain that toy will disappoint the child, but the extent of the child's unhappiness will be largely restricted to the frustrated immediate desire. Adults who fail to obtain the objects of their desires tend to have longer-lasting anger and bitterness, because failure can more severely damage their self-esteem.

In Mark's Gospel, the disciples argued among themselves about who was the greatest (Mark 9:34; see also Luke 9:46, 22:24). Jesus replied, "If any one would be first, he must be last of all and servant of all" (Mark 9:35). I think Jesus was trying to teach us that to enter the kingdom of God, we need to regard each other as equals in the eyes of God. We must remember that each of us is a child of God, and our worth derives from God's love for us.[4] When we define our worth in relation to each other, we fall into mimetic rivalries that preclude our entering the kingdom of God. The Mark passage continues, "And he took a child, and put him in the midst of them; and taking him in his arms, he said to them, 'Whoever receives one such child in my name receives me; and whoever receives me, receives not me but him who sent me' " (Mark 9:36–37).

Christianity teaches that the way to keep our focus on the core principles of love, compassion, and mercy is for all to regard each other as equally beloved children of God.

The Kingdom of God and Monotheism

What is the kingdom of God about which Jesus talked so much? Thoughtful Christians have offered a wide range of explanations, and I will share my thoughts in the hope of shedding some light.

There is a spiritual as well as a worldly component of the kingdom of God, and Jesus said "Truly, truly I say to you, unless one is born anew, he cannot see the kingdom of God" (John 3:3). I do not regard the kingdom of God as a physical place. I see it as a state of being connected to God and God's creation through faith and works. Individually, it is a state of perfect peace and contentment; collectively, it is a state of communal harmony with mutual love, caring, and respect. It is harder to experience the kingdom of God while suffering, but not impossible. Stephen seemed to be at peace with

God and the world even as he was being stoned: "And as they were stoning Stephen, he prayed, 'Lord Jesus, receive my spirit.' And he knelt down and cried in a loud voice, 'Lord, do not hold this sin against them' " (Acts 7:59–60).

As I read the Bible, I get the impression that "seeing" or "entering" or "receiving" the kingdom of God is an experience that does not lend itself to words. This is why Jesus frequently said, "The kingdom of God is like . . ." and "The kingdom of heaven is like . . ." And Jesus often used parables that generally described people doing things that involved love, caring, and respect. If the kingdom of God is a state of being at one with God, and if God has a singular essence, we will have difficulty comprehending both God and the kingdom of God, because our minds are inherently dualistic. Perhaps the reason our minds are dualistic is that we think with language, which itself is dualistic.[5] Language is dualistic because words obtain meaning from the tension between what the words *do* describe and what they *do not* describe. Words cannot describe a unitary concept that has no opposite or contrast. In other words, "big" only has meaning because it is greater than "little," and we call things by names such as chair, couch, and bed on the basis of features and functions that distinguish them from other things. Without the existence of things that are not chairs, the word "chair" would have no meaning. Consequently, our dualistic minds struggle when trying to comprehend both God and the kingdom of God—singular concepts that do not have opposites.

In addition to the difficulties posed by dualistic language, I think we also find it hard to comprehend the kingdom of God because we experience life as individuals. We do not experience what others feel, and they do not experience our feelings. Consequently, we perceive ourselves as entities distinct from the world. Among humans, dualistic language likely heightens the tendency to separate the world into two major categories—"me" and "not me." This dualism seems so natural and obvious that it is difficult to grasp Jesus' state of existence, in which the boundaries between himself and God were not distinct. Jesus said, "I and the Father are one" (John 10:30; see also 14:11). Further, Jesus blurred the distinction between himself and his disciples: "If you keep my commandments, you will abide in my love, just as I have kept my Father's commandments and abide in his love" (John 15:10). The Hebrew Scriptures seem to discourage a dualistic worldview in that God refuses to divulge God's name (Exodus 3:13–14). If God had a name, we would envision God as a distinct entity separate from the universe. The concept of the Trinity also serves to disrupt clear, distinct, dualistic boundaries, particularly because Christians generally envision the Holy Spirit as interacting with everyone (see John 14:17).

I see the kingdom of God as *both* individual and communal—our individual needs and those of the larger community are connected. If we were unified with God and God's creation, we would recognize that whatever we do to anyone or anything, we do to ourselves. I think Jesus was trying to describe the kingdom of God in terms of our relationships to God and to each other. He said, "The kingdom of God is in the midst of you" (Luke 17:21). With Jesus, relationships should be grounded on love and involve doing things for each other (John 13:14; Galatians 5:13; 3 John 1:5). Therefore, Jesus washed the disciples' feet as an act of love and humility. Before and after Jesus, prophets have also taught how God wants us to live.

The Nature of Prophets

From the perspective of Girardian theory, prophecy includes exposing the scandal of "sacred" scapegoating violence. Prophets reveal what has been hidden since the foundation of the world—that communal cohesiveness has been bought with the blood of innocent victims. Scapegoating generates a sense of camaraderie, but the social order, grounded in violence and injustice, maintains only an appearance of peace. Prophets expose as a lie the "peace" and "harmony" derived from scapegoating violence, and this is why Jesus had a prophetic voice when he said,

> You are of your father the devil, and your will is to do your
> father's desires. He was a murderer from the beginning, and has
> nothing to do with the truth, because there is no truth in him.
> When he lies, he speaks according to his own nature, for he is a
> liar and the father of lies (John 8:44).

An important feature of prophets is that they have witnessed or personally experienced scapegoating. This, I think, is why the Hebrew prophets typically had humble origins that helped them empathize with victims. Prophets recognize injustice and, if possible, they denounce it. However, they do so at great peril, because they threaten to undermine the myths, rituals, and taboos with which people orient their lives.

Jesus provided considerable insight into the nature of prophecy when he told the Pharisees and lawyers,

> Woe to you! for you build the tombs of the prophets whom your
> fathers killed. So you are witnesses and consent to the deeds of
> your fathers; for they killed them, and you build their tombs.

> Therefore also the Wisdom of God said, 'I will send them
> prophets and apostles, some of whom they will kill and perse-
> cute,' that the blood of all the prophets, shed from the foundation
> of the world, may be required of this generation, from the blood
> of Abel to the blood of Zechariah [2 Chronicles 24:21–22], who
> perished between the altar and the sanctuary (Luke 11:47–51).

The prophets condemned killing innocent victims, and many prophets, for articulating that message, were killed themselves.[6] Jesus denounced the Pharisees and lawyers for building tombs and celebrating the prophets' greatness, which made it easier for the Pharisees and lawyers to ignore the prophets' actual, challenging message.

Jesus said that the blood of all the prophets was required of this generation. I think he meant that the mindset of "this generation" was the same as that of every other generation, and therefore all generations have been equally guilty of all the murders. Each generation can condemn past murders, but it is unwilling to confront its own violence and scapegoating. Therefore many people have been angered when animal advocates have made parallels between contemporary treatment of animals and past human slavery[7] or the Holocaust.[8] They often claim that animal advocates, in elevating the importance of animals and their suffering, denigrate humans. However, animal advocates often make clear that they are showing similarities between the *mindset* of those who have victimized humans and those who currently victimize animals, as well as the similarities between the kinds of physical and psychological abuses visited upon the victims. Further, concern for animals does not require disregard for humans, just as the addition of a child to a family does not make the parents love their other children less. Indeed, harming God's animals unnecessarily degrades humans.

Prophecy

Jesus said, "No prophet is acceptable in his own country" (Luke 4:24). Some have noted that people have trouble taking seriously someone they remember as an immature youth. Gil Bailie, offering further insight, has argued that an individual becomes a prophet by virtue of being rejected.[9] The victim of ostracism (and often violence) gains an understanding of the ways in which mobs gain unity through collective violence. This is prophetic knowledge, and it requires being an outsider. To varying degrees, everyone has experienced being an outsider at times; that outsider status becomes greatly enhanced when a person identifies and objects to scape-goating, because much of what it means to be "one of us" is to agree with the rest of the community about who should be excluded due to their "evil-ness" or "inferiority." To belong to a community unified by the scapegoat-ing process requires participation in the community's scapegoating, and one's prophetic witness is therefore lost.

Those of us who are animal advocates, like others expressing their prophetic witness, aim to be "a light to the nations" (Isaiah 42:6, 49:6) through our words and actions. One consequence is that we often find our-selves alienated from our communities, because, as the ancient Hebrew prophets often experienced, people who object to the message tend to dislike the messenger. Robert C. Tannehill has written, "The destiny of God's prophets includes suffering and rejection, for they must speak God's word to a blind and resistant world and must bear the brunt of this resist-ance."[10] The scapegoating process helps explain why the world is blind and resistant. This knowledge does not make prophecy any easier or more pleas-ant, but it can help us maintain equanimity in the face of seemingly insur-mountable resistance to our message.

Because speaking with a prophetic voice can be burdensome or even dangerous, it is tempting to focus on one's own purity and righteousness and to ignore broader social injustice. However, William Sloane Coffin has written:

> Public good doesn't automatically flow from private virtue. A person's moral character, sterling though it may be, is insufficient to serve the cause of justice, which is to challenge the status quo, to try to make what's legal more moral, to speak truth to power, and to take personal or concerted action against evil, whether in personal or systemic form.[11]

Coffin speaks to prophets of all stripes. Among Christian animal protection-ists, many find that their drive to help animals is grounded in their sensitiv-ity to animal suffering. This sensitivity is a gift of the Holy Spirit that can give direction and meaning to our lives. But it is also a burden in that we often suffer empathetically with those helpless animals abused by humans, and we often find that animal activism alienates us from family and friends. With opened eyes and ears (see Mark 8:18), we recognize animals' suffer-ing and we reject the notion that victimizing them is righteous and just. In essence, we have heard the cock crow. We should not be proud or bitter about our prophetic calling—whether it seems a gift or a burden, many of us see our prophetic calling as part of God's plan.

Many people find that advocating for animals or other vulnerable, abused individuals provides a sense of satisfaction and meaning. What if we do not answer our call to prophesy? Jesus said, "Truly, I say to you, all sins will be forgiven the sons of men, and whatever blasphemies they utter; but whoever blasphemes against the Holy Spirit never has forgiveness, but is guilty of an eternal sin" (Mark 3:28–29). If we rejected our prophetic destiny, we would be committing blasphemy against the Holy Spirit.

What about those who choose to remain ignorant of the consequences of their actions, such as meat eaters who avoid learning about factory farming or people who purchase the products of slave labor? Their message to the world is not a prophetic one—their message is that it is not necessary to live by the values of their professed faith.

Will an angry and vengeful God punish those of us who disregard our prophetic destiny? I do not think so. Rather, to the degree that we reject our destiny and deny the crowing of the cock, our lives become artificial and lose integrity and meaning. Those who deny their prophetic calling are punished by their sins, not for them. By the same token, I think that prophets who abide by the Holy Spirit are rewarded by their faithfulness to God, not for it. The first challenge is to accept one's prophetic destiny; the next is to find creative ways to communicate one's prophetic witness to a resistant human community.

Prophecy and Creativity

People seem to have an innate desire to be creative. What makes one creative person's work great, and most people's writings, paintings, music, or other artistic creations ignored or quickly forgotten? Pop culture often presents human experience in simple terms, is readily accessible to a broad audience, and usually comforts people by reinforcing their values and beliefs. Pop culture has little lasting power, however, because it generally does not meaningfully describe people's greatest inner conflicts or their deepest spiritual longings and needs. In contrast, great art speaks to important aspects of human experience. The writings of the latter Hebrew prophets exemplify great literature, in part because they articulate an inspiring but challenging vision of justice and righteousness.

Often, the public resists the messages of the most insightful prophets. Communities usually reject revelation of the lies that the community wants to keep hidden, such as the lies regarding the scapegoating process. However, Jesus said, "The very stone which the builders rejected has become the head of the corner" (Luke 20:17; see also Matthew 21:42 and

Mark 12:10; reference Psalm 118:22). And occasionally, people eventually hear the prophet's message when, perhaps aided by the Holy Spirit, they are ready.

Everyone has the potential to have a prophetic voice, because everyone has experienced suffering in the form of physical pain and psychological grief, and nearly everyone has experienced being a victim of scapegoating. Nearly all of us have been falsely accused, and often there has seemed a mimetic quality to the accusation in that one person's accusation has encouraged others to join the chorus. There is often a feeling of impotent against a wave of accusations.

Even though we have first-hand knowledge about the scapegoating process, it can be difficult to recognize when we participate in scapegoating. Consequently, prophets seeking to protect victims often find it necessary to communicate subtly and indirectly, often using art forms such as fiction, poetry, painting, or music. Jesus needed to use parables to express his radical ideas to a resistant audience. I also think there is a place for prose, and I would include this book as an attempt at prophetic witness; but prose's appeal is limited to those who are ready for its message. Fiction and other art forms, being more subtle and indirect, can sometimes influence the resistant mind, but those unprepared for the prose writer's prophetic voice tend to close their minds to the message.

The creative medium used for expressing one's prophetic witness can influence whether people hear the prophetic voice. Evidently, Jesus envisioned his ministry as like a mustard tree that would grow slowly and eventually have branches for all the birds of the air (Matthew 13:31–32; Luke 13:19). Will our creative efforts bear fruit? Some will, but most will not. Paul wrote, "He who prophesies speaks to men for their upbuilding and encouragement and consolation" (1 Corinthians 14:3). Each of us has a calling to prophesy, and we need to answer that call, because without prophecy, nothing impedes injustice. The author of Proverbs wrote, "Where there is no prophecy, the people cast off restraint" (Proverbs 29:18).

Whatever we do in service to God honors and glorifies God and gives our lives meaning, purpose, and direction. When we aim to glorify ourselves with riches, sensual pleasures, and status symbols, our lives might seem pleasant to outside observers, but our sense of self-esteem often remains wanting. One reason is that such self-aggrandizement fails to address the universal human psychological and spiritual needs for a sense of meaning in life. Another is that no amount of material success can eradicate our fear of death.

Ideological Certainty versus the Quest for Truth

We desire certainty about the great existential questions, such as our origins, our post-mortem destinies, the meaning of our lives, and how we are supposed to relate to each other; but the quest for certainty can be dangerous for individuals and communities. In order to make a difference in the world, it is essential that we commit ourselves to what we believe, even to the point of great personal sacrifice. However, a sense of certainty can blind us to strategies that can help us learn, grow, and adapt to change. We need to be ready to change commitments if evidence demonstrates that our actions or beliefs have been misguided.

Ideological certainty can readily lead to injustice, because those who hold beliefs with certainty tend to resist contravening logic or evidence that might expose victimization and scapegoating. Likewise, when certainty favors one attitude or policy and compassion favors another, certainty generally overrides compassion. The typical consequence is harm and suffering.

Another problem is that, because novel perspectives might show weaknesses in a given ideology, those trying to maintain ideological certainty often have difficulty coexisting with those who hold alternative views. Ideologues often respond to conflicting perspectives by sequestering themselves in separate communities, by becoming hostile toward those who express differing views (which discourages people from challenging their beliefs), or by killing or banishing those individuals whose views threaten to "contaminate" their community.

Those manifesting ideological certainty tend to divide answers to some of the most challenging existential questions into two absolute divisions: true or false. Girardian theory indicates that such distinctions are grounded in the scapegoating process, which has generated the division between true, divinely ordained belief and false, taboo, or satanic belief. Violence, or threat of violence, maintains taboos. Commitment to ideological certainty favors rigid adherence to religious laws and often represses empathy and compassion.

Therefore, the natural desire to have certainty should not undermine the quest for truth. Recognizing the limits of our knowledge is crucial for gaining understanding about ourselves, our communities, and the nature of God, because uncertainty renders people receptive to new ideas. We need new ideas and fresh perspectives, because each of us has a very limited view of the world, and because our unconscious needs and fears can cloud our views. Sharing ideas and experiences is far more valuable and productive if we have covenantal relationships with each other. The covenants, which typically feature promises of mutual respect and commitments to truth, are often implicit rather than explicit. As Jesus said, "For where two or three are gathered in my name, there am I in the midst of them" (Matthew 18:20).

If, as Christian faith teaches, God is about love, then faith communities are severely handicapped in their attempt to understand God if they are bound by the scapegoating process. Signs that communities have been bound by scapegoating rather than their love of God include harsh, merciless, punitive laws that they attribute to God; an intolerance of "heretical" points of view; and a conviction that God loves members of their community more than the rest of God's creation. In contrast, communities guided by the faith of Christ are dedicated to love, respect, compassion, and truth.

James articulated this well: "But the wisdom from above is first pure, then peaceable, gentle, open to reason, full of mercy and good fruits, without uncertainty or insecurity. And the harvest of righteousness is sown in peace by those who make peace" (3:17–18). This is God's wisdom, and it is the wisdom that we should seek. Because humans are fallible, I am convinced that having a sense of certainty reflects only a state of mind; things we are certain about might or might not be true.[12]

The Role of Prophets

Many people find prophecy a burden, perhaps because they think that they have failed if they have not substantially reduced the injustices against which they struggle. Many of these prophets "burn out" and abandon their prophetic work. However, we are not called to save the world. We did not create the problems, and we are not obliged to fix them. Our role is to be faithful to our calling, which means doing the best we can to correct injustices, oppose victimization, and assist the afflicted. Of course, we must choose which of the many problems to address—presumably focusing on those upon which we can make the greatest impact—while always striving to avoid providing financial or moral support for any injustice.

Most of us would prefer to live comfortably and amicably in our communities rather than to "speak truth to power." However, prophets are compelled to speak and act for several reasons. First, silence in the face of injustice requires dishonesty—it means denying our values and beliefs and, in essence, being inauthentic to ourselves. Second, if we reject our prophetic destiny, we must mislead the world about our values or our actions, or both. Third, injustice generally breeds resentment, often generating a more dangerous world for everyone. Fourth, if we countenance injustice, over time it becomes easier for us to restrict our compassion and concern to fewer and fewer individuals.

Chapter 12: The New Testament and Sacrifice

This chapter looks at certain passages and themes that Christians have frequently cited to defend sacrificial violence. Although some readers of drafts of this book found this material challenging, most also related that it was helpful in explaining how adherents to a religion grounded on a ministry of love and peace have often manifested hardness of heart, violence, and scapegoating.

The Letter to the Hebrews

Some hold that The Letter to the Hebrews supports sacrificial violence, but I think a close reading suggests otherwise. The letter's author wrote, "For it is impossible that the blood of bulls and goats should take away sins" (10:4). What will take away sins, if not animal sacrifices? I want to examine Hebrews 10:8–18 closely in an attempt to answer this question. Hebrews 10:8–10 reads,

> When he said above, "Thou hast neither desired nor taken
> pleasure in sacrifices and offerings and burnt offerings and sin
> offerings" (these are offered according to the law), then he
> added, "Lo, I have come to do thy will." He abolishes the first
> in order to establish the second. And by that will we have been
> sanctified through the offering of the body of Jesus Christ once
> for all.

The writer of this passage has argued that the unsatisfactory, old sacrifices under the law have been replaced by "the offering of the body of Jesus Christ." The crucial question is this: Who made the offering? I think the text suggests that Jesus offered himself as a self-sacrifice.

Verse 10:11 reiterates that animal sacrifices cannot expiate sins: "And every priest stands daily at his service, offering repeatedly the same sacrifices, which can never take away sins." Hebrews 10:12–13 reads, "But when Christ had offered for all time a single sacrifice for sins, he sat down at the right hand of God, then to wait until his enemies should be made a stool for his feet." I think verse 10:12 describes Jesus' sacrifice as a *self-sacrifice*. This interpretation makes sense theologically, if we regard God as good. An alternative interpretation is that Jesus was sacrificed by humans, which leads to the awkward conclusion that God justified humanity via an act of scapegoating and murder. Another interpretation is that Jesus was sacrificed by God, but this interpretation would portray God as one who had killed not only an innocent man, but a man who was, according to Christian tradition, also God's beloved son.

Verse 10:13 describes Christ waiting "until his enemies should be made a stool for his feet." I see the "enemies" as the "principalities" and "powers" (Ephesians 6:12); and Jesus' way of love will ultimately triumph, because the forces of evil are self-destructive.[1] Jesus had said that Satan's attempt to cast Satan out divides the house and the house cannot stand (Mark 3:23–25; Luke 11:17–18). Jesus' followers must only wait for that to happen. This view offers a way to understand 1 Corinthians 15:24–26, which reads,

> Then comes the end, when he delivers the kingdom to God the Father after destroying every rule and every authority and power. For he must reign until he has put all his enemies under his feet. The last enemy to be destroyed is death.

Death is the ultimate enemy, because fear of death leads to scapegoating and oppressive, unjust taboos, such as those that give sacred status to "every authority and power" that rules by threat of violence. Paul described Jesus' resurrection as a triumph over death, which Jesus' followers will also experience. Whether or not resurrection refers to everlasting life following the body's demise, we can triumph over death while on earth if we experience rebirth in Christ, have a sense of affinity with the immortal God, and no longer allow fear of death to dictate our lives.

Verse 10:14 of Hebrews states, "For by a single offering he has perfected for all time those who are sanctified." Humans, inclined to sin, are not morally perfect. A Girardian understanding of this verse is that Jesus has made the single self-offering that, once made, reveals that "sacred" violence is unnecessary. Jesus has demonstrated that we are loved and forgiven by God, which means that we are "perfected" in God's eyes. How do we become "sanctified"?

Verses 15–18 clarify:

> And the Holy Spirit also bears witness to us; for after saying,
> "This is the covenant that I will make with them after those
> days, says the Lord: I will put my laws on their hearts [see
> Jeremiah 31:33], and write them on their minds," then he adds,
> "I will remember their sins and their misdeeds no more." Where
> there is forgiveness of these [sins], there is no longer any offer-
> ing for sin.

Guided by the Holy Spirit, God's laws will be on our hearts and minds. If
we embrace the love and forgiveness that these laws embody and repent of
our sinful ways, we will be sanctified. We will have no need or desire to
engage in sacrificial violence to feel sanctified.

Many Christians have interpreted Hebrews 10:8–18 to mean that
the old covenant has been replaced by a new covenant formed by the
divinely ordained sacrifice of Jesus. A reasonable alternative interpretation
is that the old sacrificial order has been abolished and replaced by a new
order, in which people have been sanctified by obedience to God. (See
Romans 12:1.) This view respects the biblical text while seeing God as cen-
tered on love rather than wrath. According to this analysis, Jesus' death was
not a sacrifice to atone for sins but rather one of many ways in which his
life was sanctified by virtue of his choosing to do God's will. What was
God's will? I think that God's will was that Jesus would help reconcile creation by taking away "the sin of the world" as mentioned in John 1:29 (i.e.,
end the killing of innocent individuals). Jesus chose the only nonviolent
way to take away the sin of the world, which was to use his teachings, life,
and death to expose "sacred" violence as scandalous.

Support for this view of Jesus' death as a *self*-sacrifice to end all sacrifices comes from the preceding chapter in Hebrews:

> Nor was it to offer himself repeatedly, as the high priest enters
> the Holy Place yearly with blood not his own; for then he would
> have had to suffer repeatedly since the foundation of the world.
> But as it is, he has appeared once for all at the end of the age to
> put away sin by the sacrifice of himself (9:25–26).

The writer of this passage says that, according to former sacrificial order,
sacrifice needed to be repeated yearly. The reason was that people needed to
regularly transfer their sins onto the scapegoat which, the writer noted, is
what had been happening since the foundation of the world. The writer also
observes that the priest shed "blood not his own"—forcing animals to suffer

the consequences of human sinfulness. The writer then points out that Jesus sacrificed *himself* in order to end all sacrifices. Similarly, the writer previously noted, "He has no need, like those high priests, to offer sacrifices daily, first for his own sins and then for those of the people; he did this once for all when he offered up himself" (Hebrews 7:27).

Earlier sacrifices involved repeated, ritual killings of unwilling victims. As discussed previously, a Girardian reading indicates that Jesus chose to accept his destiny and to sacrifice himself for all creation. Jesus did not surrender to death, but rather he chose to die to serve God's will. This view accords with John 10:18, which reads, "No one takes it [my life] from me, but I lay it down of my own accord. I have power to lay it down, and I have power to take it again; this charge I have received from my father." Jesus' *self*-sacrifice revealed the scandal of "sacred" violence, which made all future sacrificial violence not only unnecessary but also undesirable.

Michael Hardin has provided the helpful insight that in offering himself, Jesus assumed the role of the priest without consigning anyone else to become a victim.[2] Regarding Jesus' death as a self-sacrifice helps us gain a nonsacrificial understanding of Hebrews 9:22 which reads, "Indeed, under the law almost everything is purified with blood, and without the shedding of blood there is no forgiveness of sins." Keeping in mind the letter's critique of the sacrificial system, the author has *criticized* the law for appearing to justify "sacred" violence.[3] The author notes, "For it is impossible that the blood of bulls and goats should take away sins" (Hebrews 10:4; see also Hebrews 9:9, 9:12–13, 10:11). Remission of sins comes not from sacrificial violence but from forgiveness of sins (Hebrews 10:18).

What about those who have gained knowledge about sin (which from a Girardian view involves the scandal of "righteous," scapegoating violence), yet continue to participate in victimizing innocent individuals? The author of Hebrews wrote, "For if we sin deliberately after receiving the knowledge of the truth, there no longer remains a sacrifice for sins, but a fearful prospect of judgment" (Hebrews 10:26–27). With knowledge about sin, a person knows that sacrifice does not remit sin; and consequently that person is liable to harsh judgment, unlike past generations who had believed that God wanted sacrificial violence. This view accords with Jesus saying,

> Therefore I send you prophets and wise men and scribes, some of whom you will kill and crucify, and some you will scourge in your synagogues and persecute from town to town, that upon you may come all the righteous blood shed on earth, from the blood of innocent Abel to the blood of Zechariah son of Barachiah, . . . all this will come upon this generation (Matthew 23:34–36).

This generation, which has received Jesus' teachings, has no excuse for continuing to kill the prophets, as humanity has always done previously. This, I think, is why Hebrews (12:22, 24) relates,

> But you have come to Mount Zion and to the city of the living God, the heavenly Jerusalem . . . and to Jesus, the mediator of a new covenant, and to the sprinkled blood that speaks more graciously than the blood of Abel.

The blood of Abel calls for vengeance; the "more gracious" blood of Jesus calls for love and forgiveness.

Hardin has astutely related Hebrews 10:26 to Hebrews 6:1–8, which describes those who have received Jesus' message yet continue to sin, "They then commit apostasy, since they crucify the Son of God on their own account and hold him up to contempt" (6:6).[4] Although those who crucified Jesus were, like all mob participants in scapegoating, unaware that they were committing a crime, those who have heard and neglected Jesus' message not only participate in his crucifixion but hold Jesus up to contempt. Hebrews encourages us to follow Jesus' example regarding sacrifice—not to kill innocent individuals but rather to offer self-sacrifices in the form of words and deeds:

> Through him then let us continually offer up a sacrifice of praise to God, that is, the fruit of lips that acknowledge his name. Do not neglect to do good and to share what you have, for such sacrifices are pleasing to God (Hebrews 13:15–16; see also Hebrews 10:24, 12:14, 13:1–3).

As with Hebrews, many people have interpreted The Revelation to John as endorsing violence, but I think this view is mistaken.

The Revelation to John

Many people have found Revelation's apocalyptic vision appealing. It depicts destruction that will put an end to this world of suffering and lead to a better age, and this image can comfort people whose lives are filled with stress or misery. Many understand Revelation as promising ultimate victory for the select, "good" people and, equally satisfying to many people, the comeuppance for "evil" people. I have yet to meet a person of any faith who, believing in a future apocalypse, does not also believe that they are among the elect who will enjoy everlasting bliss. Another reason that many Christians have found Revelation attractive is that those bent on "holy war" in God's name claim support from Revelation's imagery.

Revelation poses a challenge for those who regard the Bible as steadily revealing the scandal of sacrificial, scapegoating violence. Revelation features many images of war and death that appear to come at the hands of God and God's forces. However, I think one can faithfully and reasonably receive Revelation in ways that accord with "God is love" (1 John 4:8, 4:16).

Evidently, Revelation's author, John the Seer, sought both to encourage those who were victims of Roman persecution and to fortify those who would likely find themselves traumatized by the anticipated conflict between God's empire and that of humanity (then represented by the Roman Empire).[5] When Revelation was written, its readers were familiar with the genre of apocalyptic literature such as that found in the Hebrew Scriptures, the Gospels, and Paul's epistles.[6] These readers likely understood that Revelation uses metaphors and should not be taken literally, as they are by some Christians today.

I find Revelation consistent with an image of Jesus as nonviolent. In Chapter 5, John the Seer describes the one who is "worthy to open the scroll and break its seals" (5:2). An elder told John to expect the Lion of Judah, a traditional symbol of military power. Instead, John the Seer wrote, "And between the throne and the four living creatures and among the elders, I saw a Lamb standing, as though it had been slain" (Revelation 5:6). People have always wanted to identify with the lion, who has control over life and death. According to a Girardian view of the Bible, God desires that we choose the lamb as our model. The lamb, who is often the victim of scapegoating violence, never victimizes anyone. Nevertheless, as discussed in chapter 10, I do not think we are obliged to accept abuse passively, and resistance can be justified.

Although there are images in Revelation that might seem to endorse violence, alternative understandings are possible and reasonable. For example,

> Now war arose in heaven, Michael and his angels fighting against the dragon; and the dragon and his angels fought, but they were defeated and there was no longer any place for them in heaven. And the great dragon was thrown down, that ancient serpent, who is called the Devil and Satan, the deceiver of the whole world — he was thrown down to the earth . . . And they have conquered him by the blood of the Lamb and the word of their testimony (12:7–9, 11).

To my reading, the "war" involved the voluntary sacrifice of the slain lamb and the testimony of his followers, who proved victorious without killing their opponents.

Revelation 19 and 20 describe the final confrontations, which many Christians have understood to depict heaven at war with satanic earthly forces. Interestingly, the "sword with which to smite the nations" (19:15) comes from the mouth of "Faithful and True," who sat upon a white horse (19:11–12). There is similar imagery later: "And the rest were slain by the sword of him who sits upon the horse, the sword that issues from his mouth" (19:21; see also 1:16, 2:12, 2:16). It is reasonable to regard the Word of God as the "sword" coming from his mouth, particularly because a sword is an image for the Word of God in Ephesians 6:17 and Hebrews 4:12. Words alone cannot commit acts of violence, but truthful words can unleash widespread violence that has been kept in check by periodic smaller doses of scapegoating violence grounded on lies. Jesus acknowledged that his teachings would disrupt families and larger communities (see Matthew 10:34–37; Luke 12:51–53), but ultimately only communities maintained by love are stable and just. Because Jesus' words undermined the scapegoating process, John the Seer anticipated great violence before the advent of peace.

Gil Bailie has noted that "apocalypse" means "unveiling." He has suggested that sanctioned violence has been veiled by religious or historical justifications that give such violence an appearance of respectability, unlike unsanctioned violence, which people call "crime." Jesus unveiled "sacred" violence, revealing that God does not want sacrifices or other kinds of violence. Once people stop regarding sanctioned violence as "justice" or "righteousness," all violence seems the same. All violence then incites reciprocal violence, conflicts escalate, and the forces of evil destroy each other.[7]

I agree with Wes Howard-Brook and Anthony Gwyther that Revelation is both descriptive and predictive. The early Christian churches faced real persecution from the Romans, and they anticipated greater struggles as they rejected the Roman Empire in favor of God's realm.[8] In addition to official sanctions from the government, the early Christians risked social ostracism. John the Seer encouraged readers to contemplate the day when righteous people will prevail after the forces of evil had destroyed themselves. Revelation 21:1–6 describes "a new Jerusalem" in which there is "a new heaven and a new earth" where "God himself will be with them; he will wipe away every tear from their eyes, and death shall be no more" (21:3–4), which accords with Isaiah 11:6–9. According to Howard-Brook and Gwyther, Revelation taught that those who resist earthly authorities and place God at the center of their world are experiencing this new Jerusalem, even if their lives remain difficult. The ultimate victory of the reign of God has been assured by the death and resurrection of Jesus.

It has seemed to me that many Christians who see Revelation as describing divinely ordained violence against satanic forces regard God as wrathful and violent. Evidently, they generally think that God demands blood sacrifices to atone for sin, and that Jesus' death was necessary to atone for the sin of the world (John 1:29). Many equate this sin of the world with Adam's original sin, which all humanity has inherited. There are problems with this theory, to which we will turn next.

Original Sin

Throughout the ages, Christians have struggled to understand why people sin, to determine the consequences of sin, and to discern how to overcome sin. A popular contemporary "atonement theology" is that everyone is sinful because everyone inherits Adam's original sin, which was Adam's disobedience to God's command not to eat the forbidden fruit. This theology holds that people must perform sacrifices to mollify God's wrath at human sinfulness (see Leviticus 4–7, 16:21–22). How can we be justified in God's eyes if we are forever tainted by "Original Sin"? The only sacrifice capable of redeeming such original sin is a perfect sacrifice, in which the victim is without sin and therefore able to carry the burden of the sin of the world. Only Jesus, who was totally innocent, could satisfy God's demand for a perfect sacrifice, making further animal sacrifices unnecessary. I will discuss difficulties with this atonement theology shortly, but here I want to look at the notion of original sin.

Augustine of Hippo (354–430) was central in developing the theory that everyone inherits Adam's original sin of disobedience. One difficulty with Augustine's theory has to do with his understanding of the mechanism of transmission of Adam's sin. Augustine maintained that human sexuality was the outward manifestation of human sinfulness (perhaps because he struggled greatly against his own sexual desires), and he asserted that the powerful passions associated with sexual intercourse transmit human sinfulness to infants.[9] With our better understanding of the biology of inheritance, Augustine's theory, or any theory that posits physical inheritance of Adam's sin of disobedience, seems unreasonable.

I think mimetic theory offers a more plausible framework for understanding original sin. Humans, as mimetic creations, inherently desire what others have or want, which strongly *predisposes* us to sin. In other words, we do inherit a *propensity* to sin, but we are not born into sin. As we grow, we develop acquisitive mimetic desires that incline us toward sin, but we still can choose whether or not to sin. All of us sin, not because we are inherently sinful, but because humans have great difficulty overcoming temptations induced by acquisitive mimetic desires. Consequently, our degree of sinfulness tends to reflect the strength of our will. Christianity can strengthen our will by offering ways to gain self-esteem that do not involve acquisitive mimetic desires and harmful mimetic rivalries.

As discussed in Chapter 2, I regard the Garden of Eden story anthropologically and as allegory, rather than historically and as literal truth.[10] I find that overwhelming scientific evidence from fields such as geology, paleontology, archeology, biology, and astronomy contradicts the literal biblical account that the universe is only about 7,500 years old. Because I am unwilling (and probably unable) to disregard well-documented scientific information, I am compelled either to receive the Genesis creation account metaphorically or to reject the story as an irrelevant fable. Taking the former approach, I interpret the story as telling us that becoming human is what inclined Adam to sin.[11]

In addition to Augustine's dubious biological explanation for the transmission of original sin, another difficulty with his theory relates to his interpretation of Romans 5:12. Augustine acknowledged that he had not mastered Greek, and some scholars have argued that Augustine, relying on an "Old Latin" version, misunderstood an ambiguously translated phrase to mean that everyone had sinned "in Adam."[12] The Revised Standard Version here reads, "Therefore, as sin came into the world through one man and death through sin, and so death spread to all men because all men sinned. . . ."

Many English translations leave unclear whether all have sinned because Adam sinned—as in the Augustinian notion of inherited original sin—or whether all have sinned because "death passed upon all men." The latter translation suggests that perennial human sinfulness is a consequence of death rather than a consequence of Adam's sin. This view accords with Becker's observations about the consequences of death anxiety on human behavior.[13]

A Girardian reading offers further insight into Romans 5:12. The sin that Adam introduced to the world was acquisitive mimetic desire that has led to harmful mimetic rivalries. Adam and Eve fell into mimetic rivalry with God because they desired the forbidden fruit that God seemed to value above all else. A consequence of the sin of acquisitive mimetic desire was death, beginning with Adam and Eve's banishment from paradise to a world of struggle and scarcity, followed by Cain's murdering Abel and the countless subsequent murders. Further, death is a cause of sin, because fear of death prompts people to try to alleviate mortality fears in at least two harmful ways. Perceiving death as punishment, people try to project their sense of guilt onto victims. By killing the victims, people temporarily feel less mortal, because the "truly guilty" individuals have received their punishment. Also, people try to feel less vulnerable to death by raising self-esteem, which they accomplish by gaining the objects of acquisitive mimetic desire. However, conflicts over the objects of acquisitive mimetic desire generate the scapegoating process.

Such efforts to relieve death anxiety are futile, because we invariably feel the progression of our bodies toward death, particularly after young adulthood, and the specter of our deaths always haunts our psyches. Though projecting guilt onto victims and dominating others to gain self-esteem might temporarily alleviate death anxiety, the relentless recurrence of mortality fears demonstrates the futility of our efforts to eradicate our fear of death. I think our best hope—perhaps our only hope—of neutralizing death anxiety is to trust in the goodness of the source of our life, which I call God. If we repeat the original sin of allowing acquisitive mimetic desires to guide our lives, we will find that our efforts fail to alleviate our fear of death. As Jesus said, "Whoever seeks to gain his life will lose it, but whoever loses his life will preserve it" (Luke 17:33; see also Matthew 16:25, Mark 8:25).

Regarding God as about life and not about death encourages us to celebrate life as a gift from God and to trust that the death of the body is not the final word. It follows that our calling is to follow and serve God. If we dedicated our lives to God, our desire to sin would fade away. Serving God includes aligning our desires with God's loving desire for all creation. I am convinced that the first Christians, who directly experienced Jesus' ministry, embraced compassion and concern for God's creation.[14] Institutional Christianity has often lost this focus. Historically, it appears that a major shift in Christian thinking and action occurred when the Christian Church aligned with the Roman Empire.

Christianity and the Roman Empire

Christianity evolved from a small Jewish movement into a major world religion, and its history has played an important role in its theology. Consequently, the relationships between early Christians and the Roman Empire have had important implications for contemporary Christian thought and belief. Although the Roman authorities initially persecuted the Christians for not worshipping the Roman emperor,[15] Constantine legalized Christianity in 313, and a series of decrees starting in 381 made Christianity the official religion of the Empire. With these changes, the Church became a powerful political force.

Among other things, these political changes profoundly influenced Christians' understanding of personal and political freedom. Elaine Pagels has noted that "the majority of Christian converts of the first four centuries regarded the proclamation of moral freedom, grounded in Genesis 1–3, as effectively synonymous with 'the gospel.'"[16] The Genesis account described God's giving Adam and Eve dominion over themselves, as well as over the rest of creation. Although God had expelled Adam and Eve from Eden after they misused their freedom, God did not strip people of free will. The early Christians held that moral freedom empowered them to control their internal passions, such as greed and sexual desire, and to resist external authorities, such as the oppressive Roman government. Gregory of Nyssa wrote,

> Preeminent among all is the fact that we are free from any
> necessity, and not in bondage to any power, but have decision
> in our own power as we please; for virtue is a voluntary thing,
> subject to no dominion. Whatever is the result of compulsion
> and force cannot be virtue.[17]

The Romans could torture and kill Christians, but the Romans could not strip Christians of their freedom to practice and believe as they chose.[18] Paul wrote,

> For freedom Christ has set us free; stand fast therefore, and do
> not submit again to a yoke of slavery. . . . For you were called
> to freedom, brethren; only do not use your freedom as an
> opportunity for the flesh, but through love be servants of one
> another (Galatians 5:1, 13).

As the Church gained political power, Christianity's emphasis on moral freedom gradually faded. In its place, Christian doctrine focused on eradicating sin, by force if necessary. Augustine's concept of original sin, which manifested itself in uncontrollable sexual desires, accorded well with this new outlook. If humans were slaves to sin, then salvation required external forces to eradicate the heretical beliefs and sinful practices to which people were drawn. In other words, Augustine's formulation of original sin provided a theological basis for a Church/Empire alliance. It seems that Jesus did not categorically oppose Roman civil authority; he said, "Render to Caesar the things that are Caesar's" (Mark 12:17; Luke 20:25; see also

Matthew 22:21). However, Jesus denounced those earthly Jewish religious authorities who were bereft of love, compassion, and mercy. I think Jesus would have decried the Church's political allegiance with Rome, which permitted repression and violence "in the name of God."

The alignment of the orthodox Christian church with the Roman Empire significantly modified Christianity's understanding of Jesus' ministry and death. As John Douglas Hall has written,

> A glorious church could not have an inglorious theology. The very idea of a faith whose central image and symbol was a crucified Jew as the official (and after Theodosius) *only* legal religion of the empire that crucified him—such an idea is absurd and to a temporal power unthinkable.[19]

The Bible depicts Jesus in a position of power at God's right hand (Matthew 26:64; Mark 16:19; Luke 22:69; Acts 2:33, 7:56; Romans 8:34), which the early Christians saw as a sign of Jesus' humility and righteousness, but which the Romans and their church allies regarded as a sign of his kingship in the stern, dominating, Roman mold.

Christians have always struggled to understand the meaning of Jesus' death, and Christendom's alliance with the Roman Empire favored theologies in which Jesus vanquished the forces of evil rather than theologies that described Jesus teaching followers how to avoid those evil forces. In other words, the Church/Roman Empire alliance encouraged theologies that regarded Jesus' death as part of a divine plan to physically conquer the forces of evil, just as the Roman Empire strove to destroy "evil" forces threatening the Empire. "Orthodox" church leaders, including Irenaeus (c. 130–202), Tertullian (c. 155–230), and Epiphanius (c. 315–403), had condemned as "evil" and "heresy" alternative understandings of Jesus' ministry; but they had few weapons other than their acerbic pens. During the fourth century, the Church started to gain the power it needed to persecute and eradicate "heretical" Christian communities and to destroy their literature.

There are many possible ways to understand the meaning of Jesus' death and resurrection, and numerous "atonement theologies" have competed with each other for the hearts and minds of Christians. Christian history helps explain why certain atonement theologies prevailed during the period of the Church/Roman Empire alliance as well as later periods, when church authorities remained aligned with reigning political forces. Girardian mimetic theory offers additional insight into the meaning of Jesus' death.

Atonement Theologies

Jesus' death was a scandal to early Christians. If he were really the son of God, and he could work miracles, how could he be humiliated, tortured, and murdered? Why did he not walk away from the cross? Christians, in struggling with these questions, have proposed various atonement theologies that suggest that Jesus died to atone for human misdeeds. I will first review leading atonement theologies,[20] which I find problematic theologically, sociologically, and politically. I will then describe J. Denny Weaver's atonement theology, which I find very appealing, in part because it is consistent with the notion that God is about love and mercy and not about violence and scapegoating.

Christus Victor (Christ the victor) was the predominant atonement theology of the early Church, and it has taken two forms. In the ransom version, the devil once held human souls in captivity. God ransomed the release of human souls by offering up Jesus as a ransom payment, and Jesus' death appeared to be a victory for the devil. But God had deceived the devil, and in raising Jesus from the dead, there was victory for Jesus and humanity.

Another *Christus Victor* theology has depicted a cosmic battle in which Jesus was killed during the battle between God and the devil. The Resurrection constituted a victory for God and placed God as ruler of the universe.

These *Christus Victor* theologies are not very popular today. The ransom theology posits that the devil had the power and authority to demand a ransom of God, which, critics have argued, would belittle God. Similarly, the cosmic battle theology uncomfortably depicts the devil having power commensurate with that of God.

In 1098, Anselm of Canterbury's *Cur Deus Homo* articulated a *satisfaction atonement* theology that has significantly influenced Christian thought for nearly a millennium. Anselm maintained that humanity's sin had offended God, and Jesus' death was necessary to satisfy God's honor. Human sin had upset the moral order, and Jesus' death was necessary to restore order to the universe. The Protestant reformers modified this theology with the notion that Jesus' death was necessary, because divine law required that sin must be punished. Humanity's sin, which relates back to Adam and Eve's original sin, had created an imbalance between good and evil in the universe, and punishment was necessary to restore order. Jesus submitted to and bore the punishment that all humans, as sinners, should have received. Many church leaders, particularly among Protestant churches, teach this *substitutionary atonement* theology as *the* explanation of Jesus' death and as an essential component of Christian faith.

The *moral influence* theology holds that Jesus' death was a loving act of God that aimed to show us that God loves humanity so much that God was willing to give up his most precious possession, his son, for humanity. This dramatic, loving act would get sinful humanity's attention and lead us toward a more righteous path.

All these atonement theologies are problematic. Suppose one asks, who killed Jesus? The first two *Christus Victor* theologies above posit the devil. However, if God permitted this to happen, then one might reasonably question God's goodness. If God tried to resist the devil but could not save Jesus from torture and death, it would appear that the devil's power approached or exceeded that of God.

The satisfaction and moral influence theologies do not blame the devil for Jesus' death, but difficulties remain. If the mob, the Roman authorities, or the high priests were responsible for killing Jesus, then one would come to the awkward conclusion that the evildoers—not Jesus—were actually fulfilling the divine mission to substitute an innocent victim for sinful humanity. However, if humans killed Jesus, it would make little sense to see Jesus' death as atonement for humanity's sins, because this would mean that sinful humanity had saved itself by killing an innocent person. In other words, murder had somehow delivered humanity from sin. Therefore, it appears that, if humanity's salvation derived from killing Jesus—whether to satisfy God's honor, to relieve humanity from the burden of original sin, or to show humanity how to live righteously—then God must be responsible. So, these theologies suggest that God either killed or orchestrated Jesus' death. This seems to portray God in an unattractive light and seems to conflict with God's previous declaration, "This is my beloved Son, with whom I am well pleased" (Matthew 3:17; see also Matthew 17:5; Mark 1:11; Luke 3:22). Further, if one believes that Jesus was God incarnate, this theology suggests that God killed God's self.

God's responsibility for Jesus' death is particularly problematic for a moral influence atonement theology. In essence, this theology holds that God engineered the death of one child, Jesus, in order to save the rest of God's children: us. Would we ever approve of a parent who had one innocent child killed in order to teach a lesson to sinful siblings?

I will focus more attention on Anselm's satisfaction atonement theology, that Jesus' death satisfied God's honor, as well as the substitutionary atonement theology that derives from satisfaction atonement theology, for two reasons: These theologies are broadly popular among contemporary Christians, and I think each creates obstacles to a Christian faith that sees God as loving and compassionate.

One problem with Anselm's satisfaction theology is that it is difficult, if not impossible, to separate God's honor from God. Therefore, Anselm's satisfaction theology seems to make the bizarre proposition that God had Jesus killed in order to satisfy God's honor.

The substitutionary atonement theology focuses not on God's honor but on the belief that all sin must be punished. The appeal of such retributive justice likely resides in the human desire for vengeance; people project their own desire to punish evildoers onto God. Substitutionary atonement theology, which attributes the need for punishment to God's desires, has several difficulties. It presumes that there are times that God wants parts of God's creation harmed; it seems incompatible with a monotheistic view of God as loving and compassionate, and it assumes that sinners are fully culpable. The last item is particularly dubious, because many factors beyond human control influence the choices people make, including their natural temperament, their early childhood experiences, the values of their parents and other adults who shaped their worldviews, and violent institutions that have unjustly deprived them of opportunities for more righteous living.

Substitutionary atonement theology has additional difficulties. It assumes that justice and righting of wrongs involve some kind of retribution. According to this framework, the problem with sin is that it causes an imbalance—a disturbance of the moral order of the universe. Given humanity's depravity and sinfulness, the only way to restore balance is through the most severe punishment: death. However, such a view separates God's justice from God's forgiveness. This separation is both theologically and socially problematic, because it encourages people to choose to either focus on God's justice or God's forgiveness, depending on their own temperament or on the moral issue at hand. When there is relative peace and well-being, people can choose to abide by the dictates of a loving and forgiving God. When there is social unrest or a crisis, people can revert to the image of God as wrathful and vengeful. Girard asserted that religions throughout history have included images of the divine as wrathful and vengeful, and this view remains attractive to many Christian communities.

Defenders of substitutionary atonement theology often point to certain biblical passages[21] and to the ancient Hebrew sacrifices[22] as proof that God desires blood sacrifices for sins. However, the Levitican sacrificial codes describe animal sacrifices for thanksgiving as well as for sins, undermining the notion that blood sacrifices fundamentally relate to punishment for sins. Weaver has noted that the Hebrews believed that the blood carried the essence of life (Genesis 9:4; Deuteronomy 12:23; Leviticus 17:11, 14).[23] The Hebrew blood sacrifices involved a priest spreading the animal's blood over the altar, where God presided.[24] Significantly, the person offering the sacrifice identified with the sacrificial animal through the laying on of hands, and consequently the lifeblood offered to God constituted a symbolic rededication of the person's entire life to God. Weaver has written, "This ritual did not involve destruction of an animal in place of killing a person. Rather, the life of the animal, namely its blood, and with it the life of the worshiper, was given to God."[25] Neither the animal's blood nor the animal's death was a penalty for human sins.

Human values, beliefs, and experiences have influenced the development of all atonement theologies. A thorough examination of any given atonement theology must include an assessment of the human context in which the theology evolved. Weaver has noted that Anselm's satisfaction atonement framework has roots in the medieval worldview. The feudal king's power resided in a belief that the king had divine authority. To offend the king was tantamount to offending God. Therefore, those who dishonored the king must be punished to restore the moral order, just as—according to contemporary satisfaction atonement theology—Jesus was killed to re-establish the moral order disturbed by original sin.

Weaver has concluded, "Any and all versions of satisfaction atonement . . . assume the violence of retribution or justice based on punishment, and depend on God-induced and God-directed violence."[26] With God involved in violence and punishment, it becomes easier for Christians to justify their own violence and punishment "in the name of God." Some might argue that, because God's wrath has been fully satisfied by the perfect sacrifice of Jesus, God is no longer wrathful. Consequently, the substitutionary atonement theology might be compatible with an image of God as one who has become purely loving, compassionate, and merciful. However, people often see themselves as victims of mistreatment, which generally incurs their wrath. It remains tempting for people who believe that they have been wronged to believe that a God of justice with the potential to become wrathful likely shares their righteous indignation. They might then regard their own vengeance as assisting God in meting out justice.

Satisfaction atonement theologies treat sin as a legal problem—humanity's offense against God—rather than as a social problem.[27] Satisfaction atonement theologies do not regard sinfulness in terms of society's institutions or events of human history (other than original sin). Consequently, satisfaction atonement theologies do not challenge unjust human institutions, making it easier for Christians to countenance injustice. With the rise of satisfaction atonement theology, Christianity's focus changed from what Jesus did and taught to what was needed to preserve Christian society. Because Christians have regarded the Church as the embodiment of God, and because religious authorities have often taught that the Church is the vehicle through which people learn the lessons about atonement that are necessary for their eternal salvation, defending the Church has often taken precedence over defending vulnerable individuals. Furthermore, there have been many times when kings and other despots have subverted the notion of Christian society to serve their own selfish desires. In such settings, the Church itself has become the world rulers in this present darkness (Ephesians 6:12) that have worked against God. Although Jesus taught that we should show love and mercy in all our relationships, satisfaction atonement theologies have changed the focus of sin from injustice against individuals to offense against God and "God's Church." Consequently, Christianity evolved into a religion that has, at various times in history, accommodated slavery, subjugation of women, cruelty to animals, and other unjust arrangements.

Some social reformers have expressed concern that substitutionary atonement theology, which is a subset of satisfaction atonement theology, facilitates neglecting victims. Substitutionary atonement theology sees Jesus' death as satisfying the penalty for sin. Now that human sin is no longer a barrier to justification before God, one may focus on one's own individual salvation and pay little attention to social justice. Although Christian doctrine generally holds that "saved" Christians naturally reflect God's love, in practice many Christians, confident of their justification before God and therefore convinced that God is guiding their moral decisions, can believe that selfish and other patently unjust behavior represents God's will.

Social reformers have pointed out another difficulty with satisfaction atonement theologies. These theologies portray Jesus as one who was innocent yet voluntarily submitted to suffering. This has often been an obstacle to people who suffer as a consequence of unjust social structures, because church authorities have often told victims of abuse, "in imitation of Christ," to submit to domestic or other abuse in the same way that Jesus accepted his tragic destiny.

Additionally, satisfaction atonement theologies are problematic in that they adopt the logic of Caiaphas who, in trying to convince the chief priests and Pharisees to call for Jesus' execution, said, "It is expedient for you that one man should die for the people, and that the whole nation should not perish" (John 11:50). Satisfaction atonement theologies posit that it is indeed better for one innocent man to die to save everyone else, which has been the logic of scapegoating violence throughout human history. Indeed, one might wonder whether satisfaction atonement theology presents Christianity as a new revelation, or whether it presents Christianity as a variation on the perennial religious theme that gods demand sacrificial violence.

Finally, satisfaction atonement theologies focus on Jesus' death; they do not require a theology about his life, teachings, or resurrection. Contemplating the meaning of Jesus' death through a Girardian lens can help us understand that God wants us to love each other and to cease scape-goating innocent individuals. We can miss this message if we divorce Jesus' death from the rest of his ministry.

Narrative *Christus Victor*

As discussed in Chapter 8, nearly universal satanic desires for power have undermined the realm of God. Revelation equates the violent and rapacious Roman Empire with Satan,[28] and countless powers and principalities have assumed this role throughout history. Provocatively, Weaver suggests that the *Christus Victor* framework lost favor "when the church came to support the world's social order, to accept the intervention of political authorities in churchly affairs, and to look to political authorities for support and protec-tion."[29] In other words, when the early church, which had no military or political clout, joined the satanic powers and principalities, it sought atone-ment theologies that removed Satan from the picture.

Although political factors might have played important roles in the declining popularity of early *Christus Victor* theologies, many early Christian theologians rejected this framework because Christus Victor posits that Satan played a necessary part in God's divine plan. However, we have seen that the satisfaction atonement theologies and the moral influence theology that have replaced *Christus Victor* theologies are them-selves problematic. In defense of *Christus Victor,* satanic forces, which I relate to acquisitive mimetic desires, are very real. They militate against the realization of God's realm in which all creation lives in peace and harmony. There is a need for humanity to atone the sins that reflect human, satanic, acquisitive mimetic desires, and Weaver has articulated an atone-ment theology that depicts Jesus' life, death, and resurrection as the means by which sinful humanity can become reconciled with God. He has sug-gested the term "narrative *Christus Victor,*" because it relies heavily on the Gospel and Revelation narratives.

Weaver's framework portrays Jesus neither as a passive victim nor as resisting his victimization by struggling to prevent the powers and princi-palities from persecuting him. If Jesus had been passive, the mob would have concluded that he agreed with their verdict that he deserved punish-ment. If he had resisted his fate, the mob, unable to see its own violence, would have seen his actions as proof of guilt. Either way, he would have been yet another victim in the endless series of scapegoats. Instead, he actively challenged the satanic structures by demonstrating, in word and deed, God's plan for universal love. He befriended and assisted a broad range of outcasts—victims of the scapegoating process. Doing so violated taboos grounded on the scapegoating process and challenged corrupt and unjust authorities, and Jesus was well aware that these actions would prompt the mob and the authorities to conspire in his torture and execution. In life, Jesus made the realm of God visible, bringing light, life, and love to the world. In death, the Resurrection established Jesus as the ultimate victor in the conflict between good and evil.

Unlike satisfaction atonement theologies, in narrative *Christus Victor*, God's role is not to engineer Jesus' death or even to tacitly condone it. The realm of God involves love, justice, and nonviolent action. This approach to relationships refuses to victimize or scapegoat, even if it leaves a person vulnerable to being a victim. Superficially, satanic powers often seem to triumph by killing peacemakers such as Jesus. However, the Resurrection marks the ultimate victory of Jesus, not Satan. In narrative *Christus Victor*, faith in the Resurrection is tantamount to faith that Jesus was innocent, that God sides with those who express love and peace rather than those who engage in violence and scapegoating, and that the realm of God will ultimately prevail. The divided house in which satanic forces try to destroy each other will eventually collapse, and creation will be reconciled to God's original intentions for a harmonious world (Mark 3:23–26; Luke 11:17–18). Christians who hold this faith can be inspired to dedicate themselves to the realm of God and, in doing so, atone for the satanic sinfulness that has gripped their lives. God forgives our sinfulness; we atone for our sins by accepting this forgiveness and aligning our desires with those of God. It is indeed by amazing grace that God forgives our participation with the same kind of powers that killed Jesus.

Jesus Made to Be Sin

Paul wrote to the church in Corinth, "For our sake he made him to be sin who knew no sin, so that in him we might become the righteousness of God" (2 Corinthians 5:21). Was it God's intention that Jesus would be sin? God made Jesus, and I think God desired that Jesus would choose to be the one whom *we humans* made into sin. Humans would heap sin upon Jesus, just as humans have heaped sin upon many scapegoats. God was responsible for making Jesus to be sin only insofar as God expected this to happen, because this is the fate of prophets. I do not think that God desired for Jesus to suffer and die; God offered Jesus this tragic destiny because God wanted to end scapegoating violence. Therefore, I regard God as involved in Jesus' death insofar as God *empowered* Jesus to expose the scapegoating process, but God did not orchestrate the Crucifixion. When Jesus defended victims and denounced victimizers, he scandalized both the Jewish and Roman authorities, making his crucifixion inevitable.

Galatians provides further insight: "Christ redeemed us from the curse of the law, having become a curse for us" (3:13). Here, Christ was a "curse," similar to the 2 Corinthians 5:21 passage in which Christ was "sin," but Galatians 3:13 additionally notes the way that occurred—through the law. Paradoxically, Paul had just written that anyone who does not keep all the law's prescriptions is cursed (Galatians 3:10). What was Paul's view of the law?

In Romans, Paul wrote that the law is "holy and just and good" (7:12) yet also the law "which promised life proved to be death to me" (7:10). How did Paul resolve this apparent contradiction? He wrote,

> Did that which is good [the law], then, bring death to me?
> By no means! It was sin, working death in me through what is
> good, in order that sin might be shown to be sin, and through the
> commandment might become sinful beyond measure. We know
> that the law is spiritual; but I am carnal, sold under sin (7:13–14).

In other words, the law is good, but human sinfulness perverts the law and makes the law an excuse for sinfulness. Indeed, Paul's own sinfulness had prompted him to use the law as an excuse for his zealous persecution of Jesus' disciples.

Therefore, the cause for Christ becoming sin was the corrupting power of sin—which comes from humanity, not God—on the law. How does Christ becoming sin allow us to "become the righteousness of God"? Paul asserted that "Christ redeemed us from the curse of the law, having become a curse for us—for it is written 'Cursed be every one who hangs on a tree'" (Galatians 3:13; see Deuteronomy 21:23). If it were God's judgment that Jesus should be cursed and deserving of crucifixion, Jesus would not have been resurrected. Therefore, sinful humanity misapplied the law, condemned Jesus as cursed, and hung Jesus on a tree (the cross).

However, the Resurrection demonstrated that the ancient Hebrew law no longer applies. Jesus had shown a new way of living. I think this view shows how one might have faith in the Resurrection while remaining agnostic about whether or not the Resurrection was a real, historical event. Faith in the Resurrection involves a belief in Jesus' innocence and a conviction that the faith of Christ accords with divine will.

A Girardian view offers additional insights. Once Jesus revealed the scandal of sacrificial violence—that the violence comes from humans and not from God—we could become righteous disciples of Jesus and servants of God. We could then receive the law as the source of the loving relationships that God intended, whereas previously the law had sometimes been used as a tool for victimizing innocent individuals.

This understanding of 2 Corinthians 5:21 provides helpful ways of looking at other passages that have seemed to favor satisfaction atonement theologies. First Corinthians 15:3 reads, "For I delivered to you as of first importance what I also received, that Christ died for our sins in accordance with the scriptures . . ." Although Paul did not clarify to which "scriptures" he referred, many people have thought that Paul was relating Jesus' death to the Levitican sacrificial code. Alternatively, one may see Jesus' death as having parallels to the Suffering Servant of Isaiah 53. This perspective, which accords with narrative *Christus Victor*, suggests that humanity's sinfulness led to Jesus' death.

Similarly, a narrative *Christus Victor* framework dovetails with a Girardian reading of 1 Peter 3:18: "For Christ also died for sins once for all, the righteous for the unrighteous, that he might bring us to God . . ." Many Jews who had regarded themselves as faithful and righteous had participated in the murder of an innocent individual, which illustrates how humanity has always been drawn to the scandal of scapegoating. Girardian insights into how and why Jesus died can help us recognize our propensity to participate in scapegoating, encourage us to reject scapegoating's attractions, and bring us closer to God.

In the final chapter, I will attempt to apply the faith of Christ to contemporary social and political issues.

Chapter 13: Contemporary Issues

Abortion

Girardian mimetic theory has offered helpful insights into the sources of many social problems. With these insights, and a determination to identify God-centered *principles* rather than definitive policies, Christians might find common ground on contentious issues that have often divided the body of Christ.

One particularly divisive issue has been abortion. In the eyes of antiabortion activists, the term "pro-choice" seems to trivialize this issue. I think this is a valid point, but I object when antiabortionists use the term "pro-life" and then limit their pro-life concerns to humans. Christianity teaches that God imbues all creatures with the spark of life and that all creation matters to God. Therefore, we should respect the God-given right to life of animals, as well as fetuses.

Though the Bible lends support to regarding abortion as the termination of human life, this does not necessarily mean that Christians should advocate making abortion illegal. Criminalizing an activity is not always the most humane, just, or efficient way to reduce or eliminate it. Other possible approaches to abortion include moral persuasion, addressing factors that lead to unwanted pregnancies, and improving social conditions in ways that make raising children more manageable.

Even if one regards protecting fetuses as extremely important, as I do, few abortion opponents regard it as the only or even the most important priority. Other considerations that can conflict with the rights of fetuses include the health of the mother, particularly if a pregnancy threatens her life; and the needs of military defense, which sometimes results in the killing of innocent civilians, including pregnant women.

Often, unwanted pregnancies result from victimization of women, including rape and incest, and sometimes men force women to have abortions. Criminalizing abortion can further victimize such women by exposing them to imprisonment and life-threatening illegal abortions. In essence, criminalizing abortion almost always scapegoats women, because men are also responsible for the pregnancy, and because men share with women responsibility for the social conditions that make many abortions seem necessary. Indeed, some women contemplate abortion, prostitution, or other unwanted paths because they live amid domestic or societal violence or injustice, and there seems to be few viable alternatives for themselves and their families.

Although difficult to gauge accurately, the rate of abortions in countries where abortion is severely restricted seems similar to that of countries where there are no such restrictions.[1] Interestingly, some of the highest abortion rates are in countries that prohibit abortions, and some of the lowest abortion rates are in countries that permit abortion. Unsafe abortions performed by non-medical personnel kill 50,000 to 100,000 women annually.[2]

We live in an imperfect world, and laws should reflect humanity's attempt to promote justice. The decision whether or not to criminalize certain activities reflects needs, values, and priorities that often conflict with each other. Therefore, attempts to address the complex personal and social issues related to abortion should involve love and compassion, and there is merit in former President Bill Clinton's dictum that abortion should be safe, legal, and rare.

Oppression of Women

Women's advocates have argued that many contemporary social institutions exploit and harm women, minorities, and other individuals. Many such women's advocates have rightly identified the injustices inherent in patriarchy, but patriarchy does not exist only to serve male privilege. Anthropologically, patriarchy has helped maintain order by defining social roles; for example, that the men hunt or earn a paycheck, while women raise the children and manage the household. Clearly defined social roles can help get needed tasks done and helps maintain a sense of social order and stability. This benefit might account for the frequency with which women embrace patriarchy and oppose women's rights.

According to Girardian theory, patriarchy, like all hierarchies, has its roots in the sacrificial process. Indeed, in Jesus' interactions with women, people with infirmities, and social outcasts, he repeatedly undermined unjust social hierarchies. Christianity does not teach that everyone has equal talent or opportunity, but everyone is equal in the eyes of God. Therefore, we should have equal consideration for each others' physical and emotional needs, and we should love our neighbors as ourselves (Leviticus 19:18; Matthew 19:19; Luke 10:27; Romans 13:9; James 2:8). Such an attitude is essential if we are to promote the kingdom of God "on earth as it is in heaven" (Matthew 6:10). Whether or not men and women assume traditional social roles, justice calls for permitting people to live outside cultural norms if they so choose, as long as they do not harm other individuals.

The Judeo-Christian revelation has played an important role in women's advocacy by encouraging people to identify with and be concerned about victims. In particular, as J.R. Hyland has noted, Jesus' ministry showed how a complete human being should manifest both archetypical male and female attributes.[3] The male principle, she has argued, involves action and overcoming. The female principle features care giving and concern. Though all people have innate desires to display degrees of both principles, cultures have often discouraged men from manifesting the female principle and women from exhibiting the male principle. However, action and overcoming without care giving and concern easily leads to violence and destructiveness; care giving and concern without action and overcoming does not prevent violence and destructiveness. Jesus repeatedly showed care giving and concern in his dealings with his disciples, friends, and even strangers; he also displayed action and overcoming, such as his defense of the adulteress (John 8:3–11), his confrontation with the heartless scribes and Pharisees (Matthew 23:13–29; Luke 11:42–43), and his turning over the money-changers' tables in the Temple (Matthew 21:12; Mark 11:15; John 2:15).

Those who embrace the male and female principles will often reflect the faith of Christ. Such people will show compassion for women as Jesus did (Matthew 9:20–23; 15:22–28; Luke 7:37–50; 13:11–13; John 4:7–26; see also parallels), as well as other vulnerable individuals. However, because of our culture's concern for victims, sometimes people succeed in victimizing others by falsely claiming to be victims.

Often, people try to determine guilt or innocence by gauging other people's intentions. However, humans are complex creatures with multiple motives influencing virtually all their decisions. Further, intentionally or unintentionally, people often mislead others regarding their motives. Finally, we often do not fully understand our own motives, much less those of other people. It is more appropriate to focus on correcting injustices than to concern ourselves about people's supposed intent.

Such an approach to social problems might seem at odds with this book's attempt to gain a better understanding of human motivations. Understanding human needs and desires is crucially important in any program designed to reduce social injustice and unnecessary suffering. Nevertheless, such general guidelines regarding human motives often do not account accurately for what motivates a particular individual in a particular setting, because each human decision reflects multiple factors that include an individual's unique life experiences. To illustrate, if we found that violence on television correlated with violence in society, we might encourage certain television programming standards. However, knowledge about the wider social effects of violence on television would not tell us why a particular person became violent at a particular time, even if that person recently viewed a violent television program.

From a practical standpoint, accusing someone of ill intent is often counterproductive. Those accused of ill intent are likely to resent the accusation, however valid it might be. Some people will modify their practices if they find that they have inadvertently harmed other individuals, but nearly everyone will become defensive and resistant to change if accused of ill intent.

Many women's advocates have focused attention on the problem of male violence against women. One source of this violence relates to men's ambivalence toward women.[4] Although men generally desire the company of women in order to address social and sexual needs, many men resent feeling emotionally and physically dependent on women. Further, in pursuit of self-esteem, many men crave exclusive access to sexual favors from attractive women. But this intense desire can make men feel vulnerable to manipulation by women, which can leave men feeling weak and humiliated. The net effect can be reduced self-esteem that breeds resentment. Men's conflicting feelings of desire and resentment can prompt aggressive behavior and violent outbursts.

Attempts by men or women to repress men's biological desires for sexual activity and men's psychological desires for self-esteem do not eradicate those desires. At best, repression displaces these desires to other objects, which can lead to obsessive behavior or scapegoating. Blaming men entirely for gender-related violence does not fully address the roots of the problem, nor does blaming women for manipulating men's desires. Often, the universal need for self-esteem contributes to conflicts between the genders. The challenge is to find a source of self-esteem that does not prompt men and women to scapegoat each other. As I have argued, adopting the faith of Christ is a Christian approach to this problem.

A final observation is that some women blame men for violence in general. Undoubtedly men perpetrate most acts of physical violence, but women contribute to violence in several ways. First, many women do participate in acts of violence, particularly against individuals weaker than themselves, including other women, children, and animals. Second, when angry at someone stronger than themselves, women can sometimes enlist men to perform violent acts on their behalf. Third, women can injure others emotionally, using words and actions to humiliate other individuals. Women need self-esteem as much as men do, though their means of gaining self-esteem and of scapegoating often differ. In conclusion, Girardian theory indicates that human mimetic desires and rivalries underlie patriarchy as well as other institutions that contribute to human violence and misery.

Social and Psychological Promiscuity

Many people have rejected religion, evidently because science now provides adequate answers to many questions about the universe, and because religious authorities have often supported unjust hierarchical power arrangements. Christian authorities have sometimes abused their power, but the spiritual founders of many of the world's religious, including Christianity, have struggled against social injustice. Indeed, it is possible that the Judeo-Christian revelation has helped us recognize the scapegoating process.

As an alternative to religion as a source of values and ethics, we have seen the rise of secular humanism, which has called for more equal consideration for all humans. However, even secular humanists have, in general, resisted applying their egalitarian principles to animals, even though humans share with many animals similar capacities to experience pain and pleasure.[5]

Commonly, secular humanists and other modern thinkers have rejected many traditional taboos as human-made and designed to facilitate exploitation. However, without taboos to delineate proper social values and behaviors, many people have found it difficult to orient their lives. This can lead to what Gil Bailey has called "psychological promiscuity"[6] in which people are receptive to a wide range of beliefs, values, and practices. Such people are mimetically influenced by friends, relatives, and celebrities quite readily, and they change their lifestyles and religions frequently. When promiscuity is seen in this light, one may regard sexual promiscuity as one possible manifestation of the broader problem of psychological promiscuity. Without traditional controls (i.e., taboos) to focus powerful human sexual desires, people can become sexually promiscuous, which tends to be both psychologically unsatisfying and socially disruptive. Similarly, as taboos against dishonoring parents diminish in favor of choosing to honor whomever one happens to respect at the moment, the family unit weakens, and familial loyalties that once helped ensure physical and psychological support become less reliable.

Nearly all of us manifest psychological promiscuity from time to time. We tend to seek new experiences and challenges, partly because we are naturally curious creatures, and partly because we have difficulty finding our need for self-esteem and worth fulfilled by our work, our activities, and our relationships. The difficulty in attaining sufficient self-esteem to quell mortality anxieties encourages a compulsive desire to find ways to feel that we are better than other people. Consequently, many of us derive pleasure from watching dysfunctional people and relationships on shows like Maury Povich and Geraldo; gossiping about the missteps and misfortunes of family, friends, and neighbors; and engaging in exploitative sexual and other relationships. However, such pleasures tend to be ephemeral and, because they cannot address the source of our anxieties, they are not fully satisfying.

With the many challenges and struggles that accompany human existence, it is often hard to follow Jesus' example of love, compassion, and mercy. Bailie has noted that we should not try to avoid the influences of other people. There is a difference between psychological promiscuity, which is largely unconscious, and consciously choosing to learn from other people. Christians need each other to discern Jesus' teachings, particularly those dealing with community.[7] As mimetic creations, we cannot help being influenced by other people. But to avoid psychological promiscuity, we need a set of mutually agreed-upon religious and behavioral standards to guide our daily decisions and long-term commitments. For Christians, Jesus' life and teachings should inform these standards, and our religious communities should reinforce them. When the mob deviates from these standards, one can more readily recognize mimetic accusation and step away from the crowd. To avoid participating in scapegoating, I suggest that the standards must deal with behaviors that harm innocent individuals and for which Judeo-Christian tradition provides unambiguous guidance. Therefore, the cases for prohibiting adultery and cruelty to animals as communal standards are stronger than the cases for prohibiting homosexuality and masturbation.

Homosexuality

Many Christians believe that the Bible unequivocally condemns homosexuality,[8] and opposition to homosexuality is a major focus of many Christian churches today. Strangely, many of those who ground their opposition to homosexuality on the Hebrew Scriptures also hold that Christians are no longer bound by the Hebrew laws, such as the kosher dietary laws. Other Christians have argued that supposedly antihomosexual passages, if understood in context, do not categorically condemn the practice.[9] Some claim that the Bible condemns promiscuity, unrestrained sexuality, and abusive sexual relationships, but not homosexuality per se.

It is remarkable that, although the Gospel narratives relate Jesus condemning a wide range of activities, there is no mention of him denouncing homosexuality. Because so many people who find themselves attracted to members of their own gender have suffered social isolation or even physical harm as a result, discerning the Bible's teachings on homosexuality is an important topic. Here, I will offer some Girardian thoughts on current controversies.

Although homosexual and heterosexual promiscuity can be socially disruptive, a loving, committed relationship between same-sex people does not victimize anyone. Given that so many other human activities clearly harm God's creation, why have Christians focused so much attention on homosexuality?

The best I can offer are theories. For one thing, homosexuality seems to violate some of the most fundamental taboos—those involving gender roles. When people define gender in sexual terms, "deviant" sexual activities would seem to risk generating social chaos. Another theory is that many people repress their feelings of same-sex attraction. They convince themselves and others that they do not harbor homosexual feelings by expressing disregard or hatred for homosexuals. A third reason for such strong antihomosexual sentiments is that homosexuals are convenient scapegoats for social disharmony, and some Christian leaders have blamed homosexuals for the breakdown of the family unit.[10] Homosexuals are a minority, they are often politically disorganized, and they have relatively few heterosexual allies. This might explain why many pastors who have railed against homosexuality have often been far less vocal about adultery and divorce, which undeniably disrupt families.

Contempt for homosexuals seems particularly inappropriate, because there is overwhelming scientific evidence that homosexuality is not a choice. Whether sexual orientation is primarily related to genetic or environmental influences, it is not chosen. Indeed, no rational person would choose a sexual orientation that frequently engenders widespread contempt and ostracism. This is likely why teenage homosexuals typically struggle with their feelings of same-sex attraction, and many relate that they felt relieved when they came to recognize and accept their homosexuality.

Many Christians who do not condemn a person for being homosexual still believe that the Bible forbids homosexual activity. Whether or not homosexual activity is sinful is a theological issue for which I find no clear biblical guidance. Unless someone is getting hurt, a person's sexual behavior is an issue between the individual and God and should not concern our Christian communities. Our Christian communities should be committed to building healthy, mutually respectful relationships. We should not exclude people because they express love and commitment to someone of the same gender—such exclusions divide the body of Christ. And it inappropriately violates the privacy of a person and that person's partner to ask about whether they have committed a sexual "sin." Consequently, I encourage churches to consider blessing lifetime commitments of same-sex people to each other. This would be an endorsement only of their commitment to each other; our churches do not need to take a stand on whether or not homosexual activity is sinful.

The Bible encourages loving, mutually supportive relationships. Because the term "marriage" has religious overtones for many people, I think secular laws should not attempt to determine whether any union should be called a marriage. I not only support the right of citizens to have same-sex civil unions, I think civil laws should treat all unions—heterosexual and homosexual—the same way, as civil unions that define the contractual legal rights and responsibilities of the two individuals.

Human Sexuality

Most people have powerful sexual desires, but human sexuality involves much more than biological drives. Acquisitive mimetic desires and rivalries strongly shape sexual desires. Because relatively few people have access to sexual activity with the most attractive partners, there is great potential for widespread injured self-esteem. The way most members of the community maintain self-esteem is by favorably comparing their own sexual partner to those whom the community mimetically regards as unattractive. In other words, because human sexuality is heavily colored by mimesis and pursuit of self-esteem, it invariably leads to social isolation and humiliation of a minority. Labeling this minority unattractive makes the bulk of the community feel good about their own sexual partners and less inclined to violently compete for the limited number of the most attractive sexual partners. In the 1960s, many people encouraged "free love" that involved relatively indiscriminate, promiscuous sex. However, many found such promiscuity unsatisfying, because the identity of sexual partners generally matters to people, and people are usually concerned about who else has sexual activity with their principal partner.

Rev. Britton Johnson has observed that we cannot be "right" about human sexuality; we can only be less wrong. He has noted,

> The mimetic forces swirling around sexuality produce all kinds of madness: objectification of attractive people; contempt for unattractive ones; competition for partners; deceptions about our motivations; perverse substitutes for interpersonal sex when rivalries become overwhelming (such as pornography, fetishes, child molestation); defining certain people as attractive and others as unattractive; economic benefits allocated according to sexual attractiveness.[11]

I regard the biblical standards about sexuality as similar to biblical rules pertaining to human violence and animal sacrifices: Many activities that the Bible permits are compromises between ideal relationships and what imperfect human beings are capable of doing. The Bible acknowledges that the best we humans can do—monogamous, committed relationships—falls short of perfection. Paul wrote, "To the unmarried and the widows I say that it is well for them to remain single as I do. But if they cannot exercise self-control, they should marry. For it is better to marry than to be aflame with passion" (1 Corinthians 7:8–9). It appears that Jesus also saw marriage as imperfect. He taught that in the Resurrection, where human relationships are presumably ideal, "they neither marry nor are given in marriage, but are like angels in heaven" (Mark 12:25).

Human relationships, particularly those in which there is sexual activity, involve restrictions and taboos that, according to Girardian theory, are grounded in the scapegoating process. Biblical standards serve to reduce, but cannot possibly eliminate, the mimetic rivalries that cause great human misery. As Johnson has noted, "Heterosexual and homosexual alike, married or celibate, we are all sexually broken."[12] As with competition for sexual partners, competition for wealth can divide communities, and Jesus gave few topics more attention than money and wealth.

Abundance versus Scarcity

Our culture, grounded in individualism and *laissez-faire* capitalism, teaches that we should regard life as a struggle to obtain scarce resources. Many people see competition for scarce resources as a zero-sum game—as one person gets more, roughly that much less is available to everyone else. There seem to be analogies in nature, in which food and other necessities are limited and animals struggle to survive and reproduce.

Jesus taught that God provides enough for everyone. He said, "Consider the ravens: they neither sow nor reap, they have neither storehouse nor barn, and yet God feeds them. Of how much more value are you than the birds!" (Luke 12:24) Does this mean that God provides enough for everyone? Everyday experience would have taught the disciples that both people and animals sometimes go hungry or even die from deprivation.

I think Jesus was trying to show that God cares about everyone, and not worrying about scarcity is important in generating loving relationships with each other and with the world at large. Our fear of physical discomfort and death prompts us to hoard essential resources, contributing substantially to scarcity. We have such concerns because we tend to put our priorities in the wrong place. Jesus said to his disciples, "Do not be anxious about your life, what you shall eat, nor about your body, what you shall put on. For life is more than food, and the body more than clothing" (Luke 12:22–23). Similarly, Jesus said, "For what does it profit a man, to gain the whole world and forfeit his life?" (Mark 8:36).

Also, Christianity teaches that material resources might be limited, but God's love for creation is not. Even though many individuals still suffer, Jesus prayed, "Thy will be done, on earth as it is in heaven" (Matthew 6:10), reflecting a dream of universal peace, harmony, and well-being.

Other passages take up this theme. Jesus said, "Take heed, and beware of all covetousness; for a man's life does not consist in the abundance of his possessions" (Luke 12:15). Then, Jesus told the parable of the rich fool who hoarded possessions, and God rebuked the man, saying that such possessions are temporary and unfulfilling (Luke 12:16–21). The Hebrew Scriptures also express this wisdom. The prophet Isaiah declared, "Why do you spend your money for that which is not bread, and your labor for that which does not satisfy?" (55:2).

Insatiable human acquisitive mimetic desires often result in scarcity. In contrast, Jesus' teachings about love and sharing assured abundance, and he said, "I came that they may have life, and have it abundantly" (John 10:10).

Wealth versus Poverty

Regarding distribution of wealth, it is remarkable that Jesus showed particular concern for poor people. Unlike the general view of his day, Jesus did not regard poverty as a sign of divine judgment. Rather, he considered poverty a consequence of human activity. Therefore, Jesus said, "As you did it to one of the least of these my brethren, you did it to me . . . as you did it not to one of the least of these, you did it not to me" (Matthew 25:40, 45).

In our society, many people seek wealth as a mimetic desire and as a hedge against the vicissitudes of fortune. However, focusing on gaining wealth distracts us from aligning our desires with those of God: Jesus said, "You cannot serve God and mammon [wealth]" (Luke 16:13). This accords with 1 John 3:17, which reads, "But if any one has the world's goods and sees his brother in need, yet closes his heart against him, how does God's love abide in him?" In 1 Timothy 6:10 it says, "For the love of money is the root of all evils; it is through this craving that some have wandered away from the faith and pierced their hearts with many pangs."

Christian faith encourages us to view the world as bountiful, certainly in terms of God's love and concern, and possibly in terms of resources. It is impossible for everyone to enjoy wealth, because wealth is a relative term. In order for some people to be wealthy, other people must be poor. However, everyone can be wealthy in a spiritual sense, and one way is to feel connected to a God of unlimited love. I am convinced that spiritual well-being addresses fundamental human needs more than material well-being, provided that basic biological needs have been met.

The ecological sciences presume that a struggle for survival is inevitable because, as Robert Thomas Malthus showed, exponential population growth invariably outstrips food supplies that, at best, increase arithmetically.[13] However, Jesus said that we should dedicate ourselves to God, not to obtaining food: "Man shall not live by bread alone, but by every word that proceeds from the mouth of God" (Matthew 4:4; see also Luke 4:4). If we align our desires with God's, we aim to help God reconcile creation to the harmonious world God intended (Genesis 1; Isaiah 11:6–9). This desire encourages us to limit our consumption and to share with others, confounding the "law" of nature that food supplies invariably become scarce.

Jesus said, "So therefore, whoever of you does not renounce all that he has cannot be my disciple" (Luke 14:33). Consequently, Jesus had these words for a young man who said he had followed the commandments and wanted to inherit eternal life:

> And Jesus looking upon him loved him, and said to him, "You lack one thing; go, sell what you have, and give to the poor, and you will have treasure in heaven; and come, follow me." At that saying his countenance fell, and he went away sorrowful; for he had great possessions (Mark 10:21–22).

I think Jesus understood that the young man's problem was that the man desired his possessions more than discipleship. There are some people who share their good fortune over time. Having material possessions, for them, is a form of stewardship rather than a means of power and control. If they were to divest themselves of everything, their long-term ability to help those in need would diminish. Whereas one may reasonably regard possessions as a means of stewardship for those in need, having possessions remains a stumbling block to the kingdom of God. Wealth substantially reduces our fear of hunger and the other hazards of poverty. Therefore, even if our aim is not to maximize our wealth, having wealth makes it more difficult for us to empathize and identify with poor people and everyone else who is weak and vulnerable. In addition, it is tempting to use wealth to satisfy our own desires rather than to address the needs of the originally intended recipients. In light of this, it makes sense that Jesus said, "How hard it will be for those who have riches to enter the kingdom of God!" (Mark 10:23)[14] This is likely one reason that the early Christian community "had all things in common" (Acts 2:44–47).

Some Christians hold that wealth is a sign of divine favor and that rich people have no obligation to assist those whose poverty reflects a moral failing. They often cite John 12:8, which reads, "The poor you always have with you, but you do not always have me." As Gil Bailie has pointed out, Jesus never asserted that poor people do not matter. Instead, Jesus was noting that his end was near, and that attending to Jesus and his legacy was important.[15]

None of us can rid the world of suffering or injustice. However, we can draw ourselves and our communities nearer to the kingdom of God when we make choices that show compassion, mercy, and love.

Freedom versus Security

Girard has noted that taboos serve to maintain social order. Consequently, "traditional values" and traditional social arrangements often have broad appeal, particularly during times of growing unrest. When there is general peace and prosperity, people feel less threatened by those who break taboos, particularly those taboos people generally regard as inessential to social order. I think that broader support for the rights of women, minorities, and homosexuals in the United States during the past few decades owes much to the country's enjoying economic growth and relative peace.[16] When people believe that enemies, either from within or without, threaten their well-being, they tend to fall back on the institutions that have provided security in the past. As discussed previously, during good times people generally ignore or ridicule prophets who advocate abandoning traditional myths, rituals, and taboos in favor of new social arrangements. During troubled times, people may violently oppose prophets and even blame the prophets for their difficulties. But it is also during times of crisis, when traditional institutions seem to be failing, that people begin to welcome the teachings of past or contemporary prophets.

Rulers have often found it convenient to convince the public that there are dangers. As long as the public does not hold the leaders responsible for the dangers, the public will tend to support those in power rather than risk the dangers inherent in adopting new social and political arrangements. However, the government can be more dangerous than the threat it purports to control. Samuel warned the ancient Hebrews that, if God granted their request for a king, the king would become a tyrant (1 Samuel 8). Because power corrupts, Jesus encouraged his followers to seek servant-hood rather than power. Jesus taught, "You know that those who are supposed to rule over the Gentiles lord it over them, and their great men exercise authority over them. But it shall not be so among you; but whoever would be great among you must be your servant" (Mark 10:42–43).

Christian faith encourages freedom and dignity for all individuals. Only extreme circumstances might warrant relinquishing some of these freedoms. If compassion, respect, and love guide our choices, and if we are "wise as serpents and innocent as doves" (Matthew 10:16), we will not stray far from a path of righteousness.

Liberal versus Conservative

Dr. Ron Leifer has offered helpful insights into what distinguishes liberals from conservatives.[17] He thinks they differ in what they regard as the source of human suffering. Liberals believe that people are basically good and that suffering is primarily due to insensitive or abusive social or legal institutions, such as the structure of the government, civil laws, and social customs. If harmful institutions could be reformed or eliminated, people would be much happier. In contrast, conservatives see human suffering as rooted in individual human failings such as laziness or moral turpitude. Conservatism, as the term is understood here, can have broad appeal because it is tied to the scapegoating process that blames individuals for troubles rather than blaming the community or its myths, rituals, or taboos.

Liberals have objected that unbridled *laissez-faire* market economies facilitate discrimination against minorities and women, engender great disparities of wealth, and leave children, animals, and the environment vulnerable to abuse. Liberals have typically favored laws to prevent abuses, while conservatives have claimed that many regulatory laws are unnecessary and harmful. Conservatives have asserted that *laissez-faire* market economies are very efficient at resource distribution and are morally just because they reward hard work, risk taking, and creativity.

According to mimetic theory, both the liberal and conservative views are narrow and flawed. The conservative position, which points to individual failings as the cause of suffering, readily lends itself to scapegoating. Even though people have widely different innate abilities, home environments, and educational opportunities, the conservative view fully blames poor people for their poverty.

Liberals generally believe that institutions have been constructed to protect privilege, but, as Rev. Johnson has astutely observed, contemporary liberalism has been heavily influenced by certain dubious assumptions related to postmodern thought.[18] Postmodernists hold that customs and moral rules, and the institutions that derive from these rules, reflect human culture. Postmodern liberals believe that they can improve social conditions by reforming or eliminating a given culture's harmful customs, moral rules, or institutions. However, postmodern liberals have generally not, as Girard has done, identified culture as the source of human desires. In contrast to Girardian thinking, postmodern liberals generally hold human desire as inherently innocent and good but corrupted by social customs, rules, and institutions. Postmodern liberals generally believe that structuring society according to supposedly good human nature will result in general well-being and perhaps even a paradise on earth.

Without addressing the acquisitive mimetic desires that generated them, can changing customs, moral rules, and institutions lead to communal well-being? In practice, many "progressive" campaigns, aiming to purge evil institutions, have done great evil themselves. People seeking to return the world to an ideal, mythic age have instigated many killing sprees. The killers have often held that their violence has been an unfortunate but necessary means to the desirable end of returning their society to an earlier, purer state. Examples of programs that used violent means purportedly to pursue utopian ends include the French Revolution, Russia's communist revolution, the Hutu massacre of Tutsi in Rwanda, and the Nazi Third Reich.[19]

Many people seem to understand intuitively that the more egalitarian programs of liberals might threaten social order and peace, which can help explain why relatively poor people often support conservative politicians even though conservatives' policies often seem to favor rich people. Successful conservative politicians generally speak to the conscious and unconscious fears of social anarchy that might accompany restructuring or dismantling institutions. Nevertheless, Jesus' curing the Gerasene demoniac (Chapter 9) did not destroy the community; and similarly we should not be afraid of rejecting unjust hierarchies grounded in the scapegoating process, particularly if we are dedicated to the love, compassion, and justice that Jesus exemplified. Women's liberation has not, as early opponents of the movement predicted, destroyed the fabric of American society; and animal liberation similarly poses little threat to general well-being.

Mimetic theory sees all violence as grounded in the scapegoating process, but people tend to distinguish between "good violence" and "bad violence." Conservatives fear "bad violence" that directly relates to mimetic rivalries. They tend to endorse "good violence" (that they generally call "defense," "national security," or "justice"), such as state-sanctioned police activities that uphold the law. Sometimes, liberals assert, this "good violence" serves primarily to maintain an oppressive order. This is most obvious in dictatorships, but it can also occur in democracies: consider the violence against those who challenged segregation laws and against women who struggled for the right to vote.[20] The liberals' fight against sanctioned violence can easily become a new form of sanctioned "good violence" that liberals call "justice," even though liberals' violence can be a mirror image of the violence they oppose.

Some people have used liberal or conservative platforms as vehicles for exploitation. Both liberals and conservatives claim to work toward just and peaceful societies, but both ideologies can be manipulated to endorse violence and victimization. Jesus taught how to generate and maintain community through love rather than by scapegoating and exclusion. Nevertheless, until our communities are infused with the Holy Spirit, we need institutions to prevent chaos. I do not regard institutions as necessarily harmful or dangerous, as long as people remember that institutions—including our church institutions—are *human* creations that require constant reevaluation and often need reform.

Animal Issues

Neither liberalism nor conservatism inherently favors animal protectionism. Both can advance or impede the work of animal defenders. Many animal advocates are liberals who see animal protectionism as a logical extension of equal consideration for humans. They argue that we should include animals within our circle of compassion for the same reasons that we include all humans, and moral consideration for humans does not depend on any mental or physical attributes. Consequently, moral consideration for animals should relate to morally relevant features, such as animals' ability to experience pain and pleasure,[21] or their being "a subject of a life."[22]

Nevertheless, most liberals have had little interest in animal issues. This also seems to be true of most liberal Christians, who have tended to fix their regard on humans.

Conservatives, as discussed above, tend to support existing institutions. Perhaps some conservatives fear that greater consideration for animals will make it more difficult to justify inequalities among people. However, some conservatives have embraced animal advocacy, notably Matthew Scully, whose book *Dominion: The Power of Man, the Suffering of Animals, and the Call to Mercy*[23] has received wide attention and praise in conservative publications. From the perspective of the scapegoating process, I see animal advocacy as potentially appealing to conservatives because animals do not seem to have sufficient moral understanding to deserve ill treatment. The view that regards animals as inferior does not lend itself well to animal rights, but it does favor animal welfare. Those who take animal welfare seriously[24] often call for dramatic changes in the way people treat animals. Such animal welfare advocates would countenance harming animals only when absolutely necessary, and they would place a high priority on minimizing animal suffering whenever harming animals is unavoidable. Unfortunately, such compassionate treatment is uncommon today, especially in Western culture.

It is possible that Christianity offers a more solid basis for animal protectionism than either liberalism or conservatism. Christian tradition holds that all life comes from God and that God cares for all creation. The Hebrew Scriptures mandate humane treatment of animals, and Jesus twice defended his healing on the Sabbath by pointing out obligations to treat animals humanely on the Sabbath (Luke 13:10–16, 14:1–5). As children of God, we should honor and obey our Creator/parent. Honoring and obeying God is incompatible with unnecessary violence and destructiveness toward anything that God has made.

Because most children have been taught with nearly every meal that killing and eating farmed animals raises no serious moral concerns, few people question the practice. Doing so would be disconcerting for several reasons: It would raise doubts about the widespread conviction that humans have the right to treat God's creation any way they desire, because it might lead to the unpleasant conclusion that one should not eat foods of which one is fond; and it would threaten core values and beliefs that have shaped the worldview of so many.

Even though animal advocates face significant resistance to their message, many people, upon learning about contemporary animal management practices, change their eating habits. Because animal exploitation industries hide their crimes against God's creatures,[25] many people do not know how much animals suffer for humans to procure animal foods, animal skins, animal experimentation data, and other alleged benefits of animal exploitation. However, those who *choose* to remain unaware, such as those who reject animal-welfare literature because they do not want to know, are rejecting God's creative goodness. Serving God faithfully requires mindfulness, and elective ignorance is no excuse.

Frequently, people justify their participation in animal cruelty by claiming that animal issues are irrelevant compared to human concerns. However, simple lifestyle choices can easily and substantially reduce a person's complicity in animal cruelty. It seems that those who decide not to take these easy steps have chosen to satisfy their personal desires for foods, clothing, and other amenities derived from animals rather than show compassion for all living beings. If this is their attitude toward animals, I wonder whether they will readily transfer this attitude to fellow humans if social, political, or economic duress threatens their lifestyles.

Christian must show love in all relationships. Paul wrote, "And if I have prophetic powers, and understand all mysteries and all knowledge, and if I have all faith, so as to remove mountains, but have not love, I am nothing" (1 Corinthians 13:2). It is easy to seem loving and compassionate when one does not feel socially, politically, or materially threatened. However, when one is put to the test, one's prophetic powers, knowledge, and faith amount to nothing without *agape*.

Although a complete review of biblical arguments for and against animal protectionism is beyond the scope of this book, many people defend human tyranny over animals by citing Genesis 1:26, in which God gives Adam dominion over the animals. Such an interpretation of "dominion" is at odds with Genesis 1:29–30, which depicts all creatures eating plants rather than each other; with the notion that Jesus gave his life to make peace and to reconcile "all things, whether on earth or in heaven" (Colossians 1:20); with Deuteronomy 17:14–20, in which God promises to choose a king for the Hebrews who will serve the people and not seek personal pleasures or wealth; and with Psalm 72, which describes how a righteous king should exercise benevolent dominion over all the lands.

Because most forms of animal abuse have become institutionalized and efficiently mechanized, and because wealthy societies can afford more luxuries such as meat and furs, it is likely that no society has ever caused more animal suffering and death than the United States does today. If animals could fully understand the magnitude of human abuse and the trivial reasons for most of it (e.g., taste preferences, fashion, and entertainment), animals would surely despise humanity. This leads me to wonder how God judges humanity, given how our society treats God's creatures.

Contemporary treatment of animals contrasts sharply with the biblical description of God's ideal in the first two chapters of Genesis. God created animals to be companions and helpers (Genesis 2:18), and Adam named the animals, which showed his benevolent relationship with them:

> So out of the ground the Lord God formed every beast of the field and every bird of the air, and brought them to the man to see what he would call them; and whatever the man called every living creature, that was its name. The man gave names to all cattle, and to the birds of the air, and to every beast of the field (2:19–20).

Note that Adam did not specifically name the species or the type of animal; he named every living creature, "every beast of the field," individually. We give individual names to those about whom we care,[26] and humans nearly always identify those animals whom humans abuse by their species name, such as "cow" or "pig." Not surprisingly, the generic names of the animals humans abuse have become epithets of contempt when applied to humans. The way we name animals today, in contrast to Adam's naming the animals, facilitates their mistreatment.[27]

Callousness toward animals has had tragic consequences for both humans and animals. Reminiscent of the Law of Karma, many of the greatest threats to our well-being relate directly or indirectly to humanity's collective hardness of heart with respect to animals. Much disease in the West results from eating animal products.[28] Animal agriculture contributes substantially to global warming, pollution, species extinctions, and other components of the growing environmental crisis, and animal agriculture squanders dwindling resources upon which our society depends.[29] Intensive animal agriculture promotes the transfer of infectious organisms from animals to people, increasing the risk of avian influenza (bird flu) and other pandemics.[30] It is difficult to gauge the spiritual cost of animal mistreatment, but it cannot increase our tendency to express empathy and compassion if we countenance abuse of innocent individuals. The spiritual disease becomes most evident when we see people deriding, belittling, or ignoring those seeking to communicate uncomfortable truths.

Regarding animal experimentation, even if it were true that the practice helps alleviate human disease, the implicit ends-justifies-the-means mindset contributes to callousness and heartlessness toward other humans. Indeed, Roberta Kalechofsky has thoroughly documented the relationship between animal experimentation and unethical human experimentation. The mindset that places knowledge above morality has led to abuses of humans throughout the world, notably by Nazi and Japanese scientists during World War II, the Tuskegee Syphilis Study in the United States from 1932-1972, and U.S. government-sponsored radiation experiments on unsuspecting citizens.[31] Contemporary safeguards have helped protect human subjects in the West, but disenfranchised people everywhere are at risk as long as our culture endorses activities that countenance abuse of innocent individuals. We know from the experiences of the last century that the cultural disease of violence is now a greater threat to human welfare than any disease of the body,[32] and a mindset that disregards animal victims can easily be extended into a disregard for human victims.

I cannot imagine that a God of love approves of humanity's massive abuse of God's creatures. As the future of humanity becomes increasingly precarious, will God grant mercy to those who have shown so little mercy toward God's animals? The Bible relates that God promised never to flood the earth again, but the Bible does not contain a promise that God will spare humanity from the consequences of its own callousness, rapaciousness, and violence.

Environmentalism and Sustainability

The psalmist wrote, "The earth is the Lord's and the fulness thereof, the world and those who dwell therein" (24:1). This is why God's instruction to Adam to till and keep the Garden of Eden (Genesis 2:15) is a sacred calling. However, humans have not been responsible stewards of God's creation.

While industrial interests generally favor "further studies" rather than action on critical environmental issues, pollution and resource depletion are clearly world problems. World temperatures are rising; land, water, and energy resources are diminishing; and species are going extinct at alarming rates. Many people troubled by these developments have favored modest lifestyle adjustments, such as driving smaller cars, recycling, and using renewable energy resources. But rarely do we hear environmentalists call for moving toward a plant-based diet.[33] This strategy may be politically wise in that it does not "scare away" meat eating people from the environmental movement, but failure to encourage a plant-based diet profoundly undermines environmentalists' campaigns.

First, animal agriculture tends to damage the environment and to significantly deplete land, water, and energy resources. Most calories and proteins are lost when farmed animals convert the plant foods they eat into flesh and other animal products, though some animals are more efficient than others at this conversion. Chickens convert about 3 pounds of grain to 1 pound of meat. The conversion ratio for swine is about 6 pounds of grain to every 1 pound of meat, and the conversion ratio for cattle is about 13 to 1.[34]

Moving toward a plant-based diet almost always reduces our footprint on the earth.[35] A major report by the United Nations Food and Agriculture Organization notes that animal agriculture is a leading cause of global warming and air pollution; land, soil, and water degradation; and biodiversity loss.[36] The report concludes that livestock contribute more to global warming than all forms of transportation combined. Farmed animals and their wastes emit huge quantities of the potent greenhouse gases methane and nitrous oxide. Because land must be cleared for grazing, animal agriculture is also a major impetus behind deforestation, especially of critical rainforests; such deforestation releases the greenhouse gas carbon dioxide and ultimately leads the planet toward desertification.

Second, those who eat animals to satisfy their taste for meat are choosing to live according to their sensual desires rather than according to environmentally friendly practices. One of the main reasons that we face a growing environmental crisis is that people have sought to satisfy their own desires for pleasure and convenience rather than abide by environmental imperatives.

Third, when people show disregard for animals' needs, they display an attitude that is spiritually dangerous for people as well as animals. The practice of selectively, and quite arbitrarily, ignoring the needs of weak and vulnerable animals makes it easier to discount or ignore the needs of people during times of stress or crisis, such as when resources appear scarce.

A Utopian Vision

What would a society grounded on Christian love look like?[37] Such a vision can help orient our goals and practices. Paul reminded his audiences that we humans have weaknesses and are inclined toward sinfulness (Romans 7:15, 7:19–20), and human sinfulness undermines any utopian program.[38] Therefore, utopianism risks self-righteousness, condescension, and projection of one's own sinfulness onto others if not tempered by humility. With these cautions in mind, I offer one of a number of possible utopian visions that would accord with Christian values and beliefs.

Members of a community guided by the faith of Christ would love each other and forgive one another's shortcomings. They would honor, respect, and celebrate individuality and diversity and would find that they could relate to each other in a much more genuine manner by not hiding their true feelings and beliefs in order to gain social acceptance. Such integrity would allow people to feel more comfortable with themselves, to have more meaningful interpersonal relationships, and to grow personally and spiritually by learning from each other. Currently, taboos tend to restrict such openness. Taboos are grounded in the scapegoating process; a culture based on Christian principles would be grounded in love.

In terms of rights, responsibilities, and laws, a society guided by Jesus' teachings would offer freedom of faith and practice, as long as those practices did not harm other individuals. It would place a high value on freedom of thought, expression, and association to prevent scapegoating and other forms of abuse by those in power. Such a society's laws would protect weak and vulnerable humans and animals, and those laws would be needed as long as some people did not have God's law of love written on their hearts. The government would exist to meet the needs of its people. Public officials would regard their positions as sacred responsibilities, and society would hold them to a high standard in terms of performing their duties effectively and honorably. A society dedicated to the principles Jesus encouraged would not seek grand wealth, making it a less attractive target for invasion.

Ideally, there would be no military, because it is always tempting for leaders to find a pretense to use a military for purposes of conquest. However, if neighboring communities were violent or aggressive, the lack of military defenses would pose a grave risk. A community grounded in and maintained by love might choose not to resist an invasion but might instead refuse to cooperate with an invading force, even though such a policy might result in brutal persecution. Such nonviolent noncooperation would require a high degree of consensus among the community's members. However, our desires for self-preservation and for the protection of family and friends, and our obligations to protect vulnerable individuals, pose great challenges to this strategy. Because there is violence and injustice in this world outside of Eden, I am not a committed pacifist. Nonetheless, I do see pacifism as a Christian ideal, and history has shown that determined people, working cooperatively and nonviolently toward an ethical ideal, can accomplish amazing things.[39]

The Challenge for Our Churches

Churches are human institutions, and as such they can easily become the principalities and powers against which Jesus struggled. If our churches are to reflect Jesus' love, compassion, and peace, they must be safe places for all God's creation. Currently, our churches often exhibit scapegoating—finding unity in their condemnation and exclusion of "nonbelievers," including those who respectfully and appropriately question church members' values or beliefs. Jesus said, "Not every one who says to me, 'Lord, Lord,' shall enter the kingdom of heaven, but he who does the will of my Father who is in heaven" (Matthew 7:21): many churches strike me as far from the kingdom of heaven.

A difficulty is that most churches must successfully market themselves to the public or they will fail to attract members. As long as people feel very free to join or leave churches, they will seek churches that offer appealing messages. In narcissistic, consumerist America, ingredients for a successful church often seem to include enthusiastic endorsement of *laissez-faire* capitalism, assurance of eternal life for true believers, and maintenance of community by scapegoating homosexuals and other marginalized individuals.

Regarding *laissez-faire* capitalism, Adam Smith showed that this economic arrangement often uses limited resources efficiently, but it can also lead to a gross disparity in wealth that leaves many people struggling in extreme poverty.

- In 1960, the world's richest 20 percent of people had 30 times more wealth than the poorest 20 percent. Four decades later, that difference had grown to 75 times more wealth for the richest.[40]

- In 1998, 1.175 billion people lived on less than $1 a day, and 2.812 billion lived on less than $2 a day.[41]

- In the 1970s, the top 100 corporate executives in the United States earned 39 times more than the average worker; today they earn 1000 times more.[42]

- The cost of family health insurance in the United States for a family of four is roughly equal to the entire income of someone who works 40 hours per week at the minimum wage. And there are about 45 million uninsured Americans, 80 percent of whom are considered the "working poor."[43]

With crumbling schools and impoverished, crime-ridden ghettos, those born into poverty, and those who otherwise find themselves disenfranchised, have little hope. Meanwhile, those with power have substantial political and social privilege, in large part because our political process has been deeply corrupted by the influence that comes from large contributions to political campaigns and parties. Our churches should be expressing outrage. Instead, they have often become a "den of robbers" (Matthew 21:13; Mark 11:17; Luke 19:46), where those with ill-gotten gains find refuge and defense. Like Jesus, our churches should be in solidarity with those who are poor and downtrodden, not with those responsible for the miserable plight of disenfranchised people.

Regarding the assurance of eternal life for true believers, many religions claim that adherence to their own specific tenets and rituals is necessary for a blissful afterlife. At most, only one of them is correct, and I suspect that many people who outwardly express confidence in their religion's validity have doubts at some level of consciousness. People often are most defensive, sometimes to the point of violence, of cherished beliefs that are vulnerable to well-reasoned argument. I think church leaders should try to help strengthen people's faith in God's love and concern to the point that they can live with uncertainty.

Regarding scapegoating, in what I regard as a complete reversal of Jesus' ministry, our churches often unify via the scapegoating process. Many church communities come together through their collective hatred of domestic "enemies," such as homosexuals, or foreign "enemies," such as the government's latest designated "evil empire." Our churches are not safe places for those labeled as "outsiders" or "unrepentant sinners," even if their "sin" harms no one. In addition, our churches are rarely safe places for God's animals. Perhaps partly because denigrating animals reinforces the conviction that humans are special creations who are particularly important to God, churches participate in abusing and tormenting God's creatures. There are churches that sponsor hunt clubs, permit hunting on their premises, and, nearly universally, regularly feed churchgoers the products of factory farming. The last offense against God's animals is perhaps the most pernicious, because it also undermines the Christian calling to spread the gospel. Paul wrote, "If food is the cause of my brother's falling, I will never eat meat, lest I cause my brother to fall" (1 Corinthians 8:13; see also Romans 14:21), and yet many ethical vegetarians find churches that sponsor the cruelty of factory farming unwelcoming places. These animal protectionists cannot countenance the cruelty behind the meals, and many have reported to the Christian Vegetarian Association that their efforts to address animal issues in Christian education or other settings have met with resistance or even hostility.

I have found widespread antipathy to Christianity among animal advocates, who understandably regard Christianity as an impediment to justice for animals rather than a venue for their liberation. Jesus liberated the animals from the Temple, but our churches rarely show more than token concern for animals who are among the most vulnerable individuals and who are arguably among "the least of these" (Matthew 25:40). How can we expect justice and peace in our communities when there is such widespread hardness of heart? How can we generate a loving, just society when Christian institutions, upon which so many people depend for guidance toward righteousness, so often reinforce this hardness of heart? Will our civilization, like Babylon (Jeremiah 50:39) and countless other past civilizations (Revelation 18:2), crumble under the weight of its arrogance, corruption, and depravity?[44] Is there any hope for the future?

Is There Hope?

> Nothing worth doing is completed in our lifetime; therefore,
> we are saved by hope. Nothing true or beautiful or good makes
> complete sense in any immediate context of history; therefore,
> we are saved by faith. Nothing we do, however virtuous, can
> be accomplished alone; therefore, we are saved by love. No
> virtuous act is quite as virtuous from the standpoint of a friend
> or foe as from our own; therefore, we are saved by the final
> form of love, which is forgiveness.
>
> Reinhold Niebuhr

Some people expect continued technological progress to provide humanity with future peace and general well-being, because science and technology have assisted humanity in many ways. Although science and technology have provided medical and material benefits, they have also produced factory farming, pollution, global warming, nuclear weapons, and sophisticated technologies of population surveillance and control. U.S. citizens have enjoyed a remarkable increase in material wealth since the 1950s, but studies indicate that Americans are more anxious and less happy than two generations ago.[45] From a global perspective, we now have sufficient resources to feed, clothe, and shelter everyone, yet poverty and malnutrition remain widespread.

Though knowledge can help us make wiser choices, it seems that our greatest hopes and fears profoundly color how we perceive and analyze reality.[46] In particular, it seems that fears related to vulnerability and death strongly influence our values and decisions. Desperate to reduce our terror, we tend to cling to violent and destructive worldviews that project our guilt, fears, and hatreds onto other individuals. As a consequence of the scapegoating process, human terror manifests itself in "sacred" myths, rituals, and taboos. Therefore, roiling beneath individual facades of psychological equanimity, and beneath cultural facades of manners and civility, are passions that readily lead to injustice and violence. Because we are often slaves to our passions, human rationality alone cannot generate justice and peace. Indeed, I have seen no association between intelligence and compassion.

Nevertheless, I believe that the human mind, if empowered by the Holy Spirit, can effectively resist the attractions of the scapegoating process. Those who have "the mind of Christ" (1 Corinthians 2:16; Philippians 2:5) or the faith of Christ (James 2:1) do not necessarily need to identify themselves as Christian or even as religious; rather, such people naturally find themselves drawn to caring, compassionate, *agape* relationships with everyone. They have a faith or worldview that addresses humanity's deep psychological hopes and fears—including a fear of death and the related fear of damaged self-esteem—and the spiritual need to connect to the source of life that Christians call God. If the members of our communities had these psychological and spiritual needs fulfilled, we would likely find ways to live and work together with respect, love, and peace, and we would strive to assure that everyone has the food, clothing, shelter, and other resources they need.

There is an urgent need for widespread adoption of the faith of Christ. For animals, human civilization without the faith of Christ has been hell on earth, manifesting itself as factory farming, recreational hunting, animal experimentation, the fur industry, rodeos, circuses, and the destruction of natural habitats. For humans, the same civilization that has provided effective medicines and surgical techniques, clean water, comfortable housing, wondrous works of art, and numerous other benefits is on a self-destructive path. Christian faith teaches that God will eventually reconcile creation (Isaiah 11:6–9), but humanity will determine the immediate future of human civilization and life on earth. In the meantime, many people live in fear of imminent catastrophe as a result of the proliferation of weapons of mass destruction; the depletion of essential land, water, and energy resources; and the growing environmental crisis.

The stumbling block is that people have always tended to assess their self-worth in relation to other people, not in relation to God. The desire to be distinct from, and better than, one's siblings, neighbors, other communities, and other countries has promoted mimetic rivalries and conflicts that have led to violence and scapegoating. If human nature compels us to participate in scapegoating and violence, there would seem to be little hope. Encouragingly, there are individuals and communities that emphasize cooperation and nurturing while discouraging violence. They offer "a light to the nations" (Isaiah 42:6; see also 49:6), confirming that people can transcend their violent tendencies.

Can powerful human motivations be channeled in non-violent directions? Ernest Becker noted that humans need self-esteem as a salve against the universal fear of death, and he was pessimistic about humanity's prognosis.[47] In general, people respond to their fear of death with behaviors that harm the earth, animals, and fellow humans. Nearly all religions, including Christianity, promise some form of personal immortality, and in theory this should assuage mortality anxieties. Nevertheless, perhaps because other religious beliefs and practices raise doubts about the veracity of one's own faith and rituals, people have tried, often violently, to eradicate the "false prophets" of other religions and the "heretical" views within their own religion. However, the Christian faith discussed in this book does not find itself threatened by alternative views. Consequently, it offers a path to self-esteem, psychological well-being, and communal cohesiveness that does not involve harming or destroying anyone else.

Will Christians and non-Christians gravitate toward a faith of Christ that favors love and compassion? Many are doubtful, because so much violence and destructiveness has been done in the name of Christ, and because many contemporary Christians seem to exhibit hardness of heart. Whether or not Christianity can inspire people to save the world from humanity's destructiveness, the faith of Christ as described in the Bible offers a path toward communal healing, as well as individual salvation. Future challenges to human civilization seem imposing, and it is unlikely that we will ever rid ourselves of physical suffering or of the anxieties related to mortality; but Jesus taught that following him can help heal broken relationships and save us from a sense of meaninglessness and despair.

If Jesus could manifest a sense of purpose and show love even while suffering and dying, we too may meet life's difficulties, including the specter of death that shadows our lives, gracefully. In other words, we can find solutions that do not make others suffer either with us or for us. Jesus taught that we should aim not to serve ourselves, but rather to serve others. Doing so allows us to participate in the world's reconciliation, rather than contribute to "the whole creation . . . groaning in travail" (Romans 8:22). This is perhaps Christianity's central claim—that God loves all creation, including each of us, and therefore God's ideal is universal peace and harmony. Consequently, faithful Christians should seek to serve God by serving others, which can give our lives direction and meaning, neutralize our fear of death, and make our lives joyful. I am convinced that this is why Jesus instructed, "If you continue in my word, you are truly my disciples, and you will know the truth, and the truth will make you free" (John 8:31–32).

• • •

Notes

Introduction

1. "Christ" comes from the Greek work *Christós*, which literally means "the anointed one." For Christians, this is Jesus of Nazareth.

2. Paraphrased from *Alice in Wonderland* by Lewis Carroll.

3. Unless otherwise stated, all translations are from the Revised Standard Version.

Chapter 1: The Scapegoating Process

1. Becker, Ernest. 1973. *The Denial of Death*. New York: Free Press.

2. Many people who have chronic diseases and have gradually prepared themselves for their own demise do not appear to experience this terror. For most of us trying to lead active, productive lives and for whom death does not offer relief from long-term suffering, knowledge of our mortality tends to be very disquieting.

3. Wolfe, Tom. 1979. *The Right Stuff*. New York: Bantam Books.

4. Research using the Terror Management Theory paradigm, discussed shortly, has indicated that fears related to stressful social situations (such as giving a speech to a large audience) or physical pain (such as dental procedures) induce anxiety, but they do not seem to generate the same need for self-esteem as death-related anxieties. Why do young children, with little or no concept of their own death, seek self-esteem? Fears of injury are among the greatest concerns of young children, and self-esteem gives them confidence to play and explore. Whether concerns other than death contribute to adult humans' strong need for self-esteem is not crucial to this book's main points; what is central is that humans crave self-esteem.

5. Becker, Ernest. 1971. *Birth and Death of Meaning, 2nd Ed*. New York: The Free Press.

6. Solomon, Sheldon, Jeff Greenberg, and Tom Pyszczynski. 1998. "Tales from the crypt: On the role of death in life." Zygon vol. 33, pp. 9–44; Pyszczynski, Tom, Sheldon Solomon, and Jeff Greenberg. 2003. *In the Wake of 9/11: The Psychology of Terror*. Washington, DC, American Psychological Association.

7. In what might appear to be contradictory evidence, it is my understanding that some forms of meditation, including that of Tibetan Buddhism, begin with a meditation on the inevitability and unpredictability of death. Evidently, this meditation helps relieve the natural human desire to grasp at life and cling to worldly things. I suspect that, in contrast to Tibetan Buddhists

and similarly minded people, most of us try to suppress thoughts about death, because we tend to find thinking about death unpleasant.

TMT studies have shown that reminders of mortality are disquieting, and subjects respond with behaviors aimed to reduce mortality-related anxieties. In contrast, it appears that prolonged and disciplined meditation on death eventually makes thinking about death compatible with peace of mind. Eradicating the need to repress thoughts about mortality can be spiritually liberating.

8. Harmon-Jones, Eddie, Linda Simon, Jeff Greenberg, Tom Pyszczynski, Sheldon Solomon, and Holly McGregor. 1997. "Terror management theory and self-esteem: Evidence that increased self-esteem reduces mortality salience effects." *Journal of Personality and Social Psychology* vol. 72, pp. 24–36.

9. Pyszcznski, et al., *In the Wake of 9/11*, note 6, pp. 74–77.

10. Becker, Ernest. 1971. *The Birth and Death of Meaning*, 2nd ed. New York: Free Press, pp. 68–69.

11. *Ibid*, p. 70.

12. Tavris, Carol, and Elliot Aronson. 2007. *Mistakes Were Made (but not by me): Why We Justify Foolish Beliefs, Bad Decisions, and Hurtful Acts.* New York: Harcourt, pp. 192–195.

13. *Ibid*, pp. 79–80.

14. Sometimes political leaders cynically scapegoat innocent individuals to shift blame for a failure or crisis from themselves onto other individuals. The masses must believe in the victim's guilt, or they will remain resentful of the political leaders and thirst for vengeance.

15. Girard, René. 1972. *Violence and the Sacred.* Baltimore: Johns Hopkins University Press. For our purposes, it is important to note that in primal societies, ritual sacrifice renews the sense of camaraderie that the original sacrifice engendered. In the next chapter, we will see the role of scapegoating violence in the Hebrew Scriptures.

16. Gadalla, Moustafa. 1999. *Exiled Egyptians: The Heart of Africa.* Greensboro, NC: Tehuti Research Foundation.

17. There are some traditions that reject the human desire to participate in mimetic rivalry or to engage in vengeance, most notably Buddhism. However, as Britton Johnson notes in an article accessed on the Internet in 2005 but that is not currently available, this reflects a conscious effort to expunge the human desire to participate in mimetic rivalries and does not necessarily refute the claim that the culture is founded on sacred violence.

18. Girard, René. 1978. *Things Hidden Since the Foundation of the World.* Stanford: Stanford University Press.

19. I acknowledge David Sztybel for important insights in developing the ideas of this paragraph.

20. Nuechterlein, Paul J. 2002. René Girard: "The Anthropology of the Cross as Alternative to Post-Modern Literary Criticism." Last accessed April 5, 2008 from http://girardianlectionary.net/girard_postmodern_literary_criticism.htm.

21. Hamerton-Kelley, Robert G. "A Religious Anthropology of Violence: The Theory of René Girard" [talk at Salt Lake City, 10/28/02].

22. Girard, René. 1986. *The Scapegoat.* Baltimore: Johns Hopkins University Press.

23. 1 Samuel 15:22; Isaiah 66:3; Jeremiah 6:20, 7:22; Hosea 6:6; Amos 5:21–22; Micah 6:6–8.

24. Akers, Keith. 2000. *The Lost Religion of Jesus: Simple Living and Nonviolence in Early Christianity.* New York: Lantern Books; Phelps, Norm. 2007. *The Longest Struggle: Animal Advocacy from Pythagoras to PETA.* New York: Lantern; Nelson, Walter Henry. 1996. Buddha: His Life and His Teaching. New York, Penguin Putnam; "Dharma Data: The Caste System", in Malalasekera, G.P. and Jayatilleke, K.N. 1968. Buddhism and the Race Question UNESCO, 1968. Available on the Internet at www.buddhanet.net/e-learning/dharmadata/fdd53.htm, last accessed April 5, 2008.

25. "Myth" as I use it here does not imply that it is a false story. As Mary Midgley has written, "Myths are not lies. Nor are they detached stories. They are imaginative patterns, networks of powerful symbols that suggest particular ways of interpreting the world." See Midgely, Mary. 2003. *The Myths We Live By.* New York: Routledge, p. 1.

26. Animals care about their well-being in ways similar to humans in terms of having enough food, avoiding extremes of weather, and seeking companionship to degrees typical of their species. Further, many animals are deeply invested emotionally in raising young and protecting their friends, who may or may not be members of their own species. See Bekoff, Marc and Dale Jamieson, eds. 1990. *Interpretation and Explanation in the Study of Animal Behavior, Volume 1: Interpretation, Intentionality, and Communication.* Boulder, CO: Westview Press.

27. Fiddes, Nick. 1991. *Meat: A Natural Symbol.* New York: Routledge.

28. Williams, Erin E., and Margo Demello. 2007. *Why Animals Matter: The Case for Animal Protection.* Amherst, NY: Prometheus Books; Nation Earth Organization. 2006. Earthlings [videotape]. Burbank, CA; Pollan, Michael. 2006. *The Omnivore's Dilemma: A Natural History of Four Meals.* New York, NY: Penguin Press. Approximately 10 billion land animals are killed for food in the United States each year. Laboratory

researchers kill roughly 30 million animals. About 20 million animals are killed for furs, and the pain of the leghold trap and the misery of the fur farm are among the most horrific examples of animal cruelty. Probably the greatest amount of suffering occurs on factory farms. Animals routinely experience painful mutilations without painkillers, stressful crowding, and chronic frustration due to their inability to express instinctive drives. The practice of tail-docking pigs illustrates the inherent cruelty of factory farming. Farmers remove piglets from their mothers at ten days, so the pigs retain a life-long desire to suck and chew. If given the chance, they would bite neighboring pigs' tails. Normal pigs would fight off such attacks, but stressed, chronically depressed pigs often fail to fight back. Their skin could become infected, which would cost money to treat, leaving farmers with a choice to absorb this expense or kill infected pigs. To stop pigs from engaging in this tail-biting "vice," the U.S. Department of Agriculture has recommended that farmers cut off most of the piglets' tail, a procedure done without painkillers, often leaving a small stump, which is very sensitive. Consequently, if another pig tries to bite a pig's tail stump, the pain is so intense that the pig will fight back, reducing the risk of an infection. While such a strategy to avoid this consequence of factory farming strikes me as diabolical, it makes perfect economic sense, as long as the pain and suffering of pigs have no value. (See *The Omnivore's Dilemma*, p. 218.)

29. Midgley, Mary. 1995. *Beast and Man: The Roots of Human Nature.* London: Routledge; Midgley, Mary. 1984. *Animals and Why They Matter.* Athens, GA: University of Georgia Press.

30. Griffin, Donald R. 1992. *Animal Minds.* Chicago: The University of Chicago Press.

31. Bekoff, Marc. 2007. *The Emotional Lives of Animals.* Navato, CA: New World Library; Masson, J. Moussaieff. 2003. *The Pig Who Sang to the Moon: The Emotional World of Farm Animals.* New York: Ballantine Books; Masson, Jeffrey Moussaieff, and Susan McCarthy. 1995. *When Elephants Weep: The Emotional Lives of Animals.* New York: Delacorte Press.

32. Rose, Margaret and David Adams. "Evidence for pain and suffering in other animals," in Langley, Gill, ed. 1989. *Animal Experimentation: The Consensus Changes.* New York: Routledge, Chapman and Hall, pp. 42–71; Grandin, Temple, and Catherine Johnson. 2005. *Animals in Translation: Using the Mysteries of Autism to Decode Animal Behavior.* New York: Scribner.

33. Chervova, L.S. 1997. "Behavioral reactions of fish to painful stimuli." *Journal of Ichthyology.* Retrieved from the Internet April 5, 2008 at

http://images.nature.web.ru/nature/2003/10/02/0001195504/english.html; Chandroo, K.P., I.J.H. Duncan, and R.D. Moccia. 2004. "Can fish suffer? Perspectives on sentience, pain, fear, and stress." *Applied Animal Behavior Science* vol. 86, pp. 225–250.

34. Phillips, Mary T. 1991. *Constructing Laboratory Animals: An Ethnographic Study in the Sociology of Science* [Ph.D. thesis]. New York University, pp. 177–220; Arluke, Arnold B. 1988. Sacrificial symbolism in animal experimentation: Object or pet? *Anthrozoös* vol. 2, pp. 98–117.

35. Barnard, Neal. 1993. *Food for Life*. New York, NY: Harmony Books; Eisman, George. 2006. *A Basic Course in Vegetarian and Vegan Nutrition*. Burdett, NY, Diet-Ethics Publishing; McDougall, John A. and Mary A. McDougall. 1983. The McDougall Plan. Piscataway, NJ: New Century Publishers. To illustrate how modern factory farming generates unhealthy foods, consider the ratio of omega-3 essential fatty acids (predominantly produced by plant leaves) to omega-6 essential fatty acids (predominantly produced by plant seeds), which is roughly 1 to 1 among hunter-gatherers. A higher ratio of omega-6 fatty acids can contribute to heart disease, and the ratio in the modern industrial diet is more than 10 to 1, thanks in large part to a shift in animal feed from leaves to seeds, primarily corn. Farmed salmon consume large amounts of corn and their flesh might have an even higher ratio of omega 6 to omega 3 than grass-fed cows. See *The Omnivore's Dilemma*, pp. 267–269.

36. Luke, Brian. 1996. "Animal experimentation as blood sacrifice." *International Society for Anthrozoology Newsletter* vol. 12 (no. 11), pp. 4–7.

37. Sometimes, researchers assert that animals benefit from animal experiments, particularly veterinary research. What is remarkable about this claim is that it denies that animals used in experiments have importance as unique individuals. We would certainly regard the possible medical benefits of killing humans in experiments as an inadequate excuse for the practice. Actually, it does not appear that utilitarian calculations — benefits for animals outweighing the costs for animals — dictate veterinary research. The choice of which veterinary research projects to pursue seems to be heavily influenced by financial considerations, such as the development of profitable drugs and treatments.

38. National Trappers Association. Undated. *NTA Trapping Handbook*. Sutton, NE: Spearman, p. 4.

Chapter 2: The Hebrew Scriptures

1. Sproul, R.C. 2005. *Scripture Alone: The Evangelical Doctrine.* Phillipsburg, NJ, P&R Publications; Evans, Anthony T. 2004. *The Transforming Word: Discovering the Power and Provision of the Bible.* Chicago, Moody Publishers; Snow, Eric V. Is the Bible the Word of God? Last accessed April 5, 2008 from www.rae.org/bibref.html.

2. Burr, William Henry. 1987. *Self-Contradictions of the Bible.* New York: A. J. Davis & Company (original publication 1860); Ehrman, Bart D. 2000. *The New Testament: A Historical Introduction to the Early Christian Writings.* New York: Oxford University Press; Ehrman, Bart D. 2007. *Misquoting Jesus: The Story Behind Who Changed the Bible and Why.* New York: HarperCollins.

3. Wroe, Ann. 1999. *Pontius Pilate.* New York: Random House. I thank Robert Kalechofsky for helpful insights on this point.

4. I acknowledge Rev. Lisa Hadler for this observation.

5. Gamble, Harry Y. 1985. *The New Testament Canon: Its Making and Meaning.* Philadelphia: Fortress Press.

6. The Bible is comprised of words, and words do not have absolute "inerrant" meanings. Instead, words have different meanings to different people. In learning language, which is a lifelong activity, we learn what words mean by observing how others use them. Because each of us has different experiences as we learn language, our understandings of words' meanings can differ subtly or markedly from others' understandings. Each of us has a unique language, but our use of words is sufficiently similar to allow effective communication of ideas. Still, misunderstandings are common, particularly among people with different social backgrounds. Therefore, each of us will receive the Bible's words in slightly different ways, and no one way is exclusively "correct," including a reading of the Bible in the original Hebrew and Greek. Even if God directed the Bible's authors, nobody can claim to understand what God intended to communicate, because each us receives the Bible's words with our own, unique language, which means that everyone receives the Bible's words differently. Adding to this language difficulty, there is considerable uncertainty about the meaning of many ancient Hebrew and Greek words, particularly seldom-used words, and some of the Hebrew and Greek words do not have close analogues in the English language. Biblical translations must rely on analyses of how the ancient Hebrew and Greek words were used in different places in the canonized literature and noncanonical texts, and the limits of human knowledge limit the accuracy of translations. Consequently, there are often multiple reasonable but different readings of certain passages.

    The New Testament contains further language difficulties. Jesus spoke in Aramaic, and his sayings were related by listeners whose memories were likely faulty, at least in details. Along with the inherent difficulties associated with translations, the original Aramaic sayings of Jesus and his disciples were translated into Greek, the original language of the New Testament. No original Gospels remain, and scholars dispute the relative authenticity of copies that contain differences among them. For example, scholars generally agree that the earliest Gospels do not contain Mark 16:9–20, which includes Jesus claiming that believers will speak in tongues, handle serpents, and cure diseases by laying on their hands (see Ehrman, *op. cit.* note 2). In fact, there are more points of disagreement among ancient copies of the Gospels than there are words in the Gospels.

7. Another difficulty relates to problems inherent in our interpreting the Bible's sayings and stories. Nobody hears or reads a story without beliefs, values, experiences, and preconceived notions that color how they interpret the story. For example, when pro-Israeli and pro-Palestinian observers viewed media coverage of the Israeli invasion of Lebanon in the early 1980s, both groups strongly believed that the coverage was biased heavily toward the other side. (See Vallone, Robert P., Lee Ross, and Mark R. Lepper. 1985. "The hostile media phenomenon: Biased perception and perceptions of media bias in coverage of the Beirut massacre." *Journal of Personality and Social Psychology* 49:577–585; and Perloff, Richard M. 1989. "Ego-involvement and the third person effect of televised news coverage." *Communication Research* 16:236–262.) Preconceived values and prejudices help explain how slaveowners once interpreted certain biblical passages as endorsing slavery, but today nearly all Christians interpret these same passages very differently.

   A related difficulty is that no matter how much translators have attempted to be accurate, their biases invariably influence their work. There are many Hebrew and ancient Greek words whose meanings are unclear or which do not translate readily into other tongues. Here, translators' views of what they think the Bible is trying to say invariably affect the translations. Those who compare English translations will find important differences in certain key passages, and these differences have had broad implications for Christian theology and have sometimes led to schisms within Christendom.

8. Whatever animals were like before the time when humans gained self-consciousness, scientific investigations—primarily studies of animals in their natural habitats—have found that many animals today exhibit evidence of thoughts, feelings, and self-consciousness. See Bekoff, Marc. 2006. *Animal Passions and Beastly Virtues*. Philadelphia: Temple University Press; and Masson, Jeffrey Moussaieff. 1995. *When Elephants Weep: The Emotional Lives of Animals*. New York: Delacorte; Midgley, Mary. "Persons and Non-

Persons," in Singer, Peter, ed. 1985. *In Defense of Animals*. New York: Basil Blackwell, pp. 52–62.

9. There is certainly archeological evidence that creatures prior to Homo sapiens possessed human self-consciousness. For example, the elaborate death rituals of the Neanderthals indicate that they anticipated that their sense of self would participate in some kind of afterlife. Further, there is scientific evidence of empathy in some more cognitively skilled animals. This not only illustrates moral character among animals but also suggests an important degree of abstract thinking ability. In one study, rhesus monkeys, upon pulling one of two chains to get food, observed through a one-way mirror a second rhesus monkey receiving a simultaneous electric shock. One chain caused a fellow monkey to receive an electric shock, and the other did not. Ten of fifteen monkeys preferred the non-shock chain, and two monkeys did not pull either chain, preferring instead to go without food for 5 days and 12 days. Self-starvation was more likely among monkeys who had previously received electric shocks themselves. (See Masserman, Jules H., Stanley Wechkin, and William Terris. 1964. " 'Altruistic' behavior in rhesus monkeys." *American Journal of Psychiatry* vol. 121, pp. 584–585.) The documentary *People of the Forest* depicts an adult male chimpanzee who watched over and protected an unrelated, crippled, adolescent chimpanzee from the torment of other adolescent chimpanzees. (See Van Lawick, Hugo, director. 1991. *People of the Forest*, Discovery Channel Video.)

10. Girard, René. 1979 [1972]. *Violence and the Sacred*. Baltimore: Johns Hopkins University Press, p. 4.

11. It has been my impression, based on experiences leafleting and tabling, as well as reading Internet messages to the Christian Vegetarian Association, that this is the most frequently cited passage among those aiming to defend meat eating.

12. Williams, James G. 1991. *The Bible, Violence, and the Sacred*. Valley Forge, PA: Trinity Press, pp. 46–50.

13. *Ibid*, pp. 29–54.

14. Girard, René. 1977. *Violence and the Sacred*. Baltimore: Johns Hopkins University Press, pp. 56–59, 61–63.

15. In many biblical stories, parents augment sibling rivalries by favoring one child.

16. Alison, James. 2003. "Violence & Christian Faith," in *James Alison in Santa Fe* [audiotape], August 22.

17. Miles, Jack. 1995. *God: A Biography*. New York: Alfred A. Knopf, pp. 314–328.

18. *Ibid*, p. 323.

19. *Ibid*, pp. 318–323.

20. *Ibid*.

21. Schwager, Raymund. 2000 [original publication in German 1978]. *Must There be Scapegoats? Violence and Redemption in the Bible*. New York: Crossroad Publishing, pp. 60–61.

22. *Ibid*, p. 61.

23. *Ibid*, p. 93.

24. Williams, note 12, pp.71–103.

25. *Ibid*, p. 89.

26. Although many Christians believe that Jesus assumed the role of the Suffering Servant, and that we are called to emulate that role, I do not think the Suffering Servant story was written by the ancient Hebrews to predict Jesus' ministry. For one thing, to my reading, the Hebrew Scriptures do not indicate that the Suffering Servant is a future personage. For another, Isaiah 53:2–3 describes the Suffering Servant as ugly and friendless, which does not accord with New Testament accounts of Jesus. I see the Hebrew writings, particularly those of the later prophets, as remarkably insightful about the scapegoating process; they describe how to live righteously and faithfully. Therefore, I have no quarrel with those who hold that the Hebrew Scriptures are an adequate revelation of God's will. Jesus said, "Think not that I have come to abolish the law and the prophets. I have come not to abolish them but to fulfil them" (Matthew 5:17). It follows that Jews who understand the law and the prophets understand God.

27. Some scholars maintain that the Suffering Servant was meant to represent the nation of Israel. The Servant's suffering relates to Israel's experience when most of its people were exiled in Babylon. Whether the Suffering Servant was an individual or a nation, the text describes the scapegoating process.

28. The *New International Version* is distinctive in translating Jeremiah 7:22 as follows: "For when I brought your forefathers out of Egypt and spoke to them, I did not just give them commands about burnt offerings and sacrifices." Adding the word "just" completely changes the passage's meaning.

29. Buttrick, George Arthur, ed. 1980. *The Interpreter's Bible, Volume 1*. Abingdon: Nashville, p. 983.

30. I thank Rev. Frank L. Hoffman for helpful insights here.

31. Akers, Keith. 2000. *The Lost Religion of Jesus: Simple Living and Nonviolence in Early Christianity*. New York: Lantern Books.

32. Shoeps, Hans-Joachim. 1964. *Jewish Christianity: Factional Disputes in the Early Church*. Philadelphia: Fortress Press; see Akers, note 31.

33. I acknowledge Rev. Paul J. Nuechterlein for insights here.

Chapter 3: The Life and Death of Jesus

1. Girard, René. 2001. *I See Satan Fall Like Lightning*. Maryknoll, NY: Orbis Books, 1991; Heim, S. Mark. *Saved from Sacrifice: A Theology of the Cross*. Grand Rapids, MI: Wm. B. Eerdmans Publishing, 2006.

2. Nuechterlein, Paul J. 2004. "The Baptism of Our Lord." Last accessed April 5, 2008 from http://girardianlectionary.net/year_c/epiphany1c.htm.

3. Campbell, Joseph. 1972. *The Hero with a Thousand Faces*. Princeton: Princeton University Press.

4. Borg, Marcus J. and John Dominic Crossan. 2006. *The Last Week*. San Francisco: HarperSanFrancisco.

5. Scholars have noted that the Synoptic Gospels (Matthew, Mark, and Luke) identify disputes between Jesus and particular Jewish groups and authorities, notably the Pharisees, Sadducees, scribes, and high priests. These Gospels were written when many Christians retained a Jewish identity and regarded themselves in relation to other specific Jewish groups. By the time John's Gospel was written, the Jewish Christian movement had been overwhelmed by Gentile Christianity. There was little reason for the community for whom John wrote to distinguish among the various Jewish groups that had opposed Jesus' ministry. Consequently, John's Gospel describes Jesus' Jewish opponents simply as "the Jews." Sadly, the Passion narratives have often helped fuel anti-Semitism, making it easier to scapegoat Jewish people.

6. Bailie, Gil. 1997. *Violence Unveiled: Humanity at the Crossroads*. New York: Crossroad Publishing, p. 218. Here, "all" refers to the people present at the Crucifixion.

7. Girard, René. 1978. *Things Hidden Since the Foundation of the World*. Stanford: Stanford University Press.

8. Funk, Robert W. and The Jesus Seminar. *The Acts of Jesus: What Did Jesus Really Do?* 1998. New York: HarperCollins; Erdman, Bart D. The New Testament, Lecture 1 [audiotape series]. 2000. Chantilly, VA: The Teaching Company.

9. Armstrong, Karen. 2004. *The Spiral Staircase: My Climb out of Darkness*. New York: Knopf.

10. Girard, René. 1972. *Violence and the Sacred*. Baltimore: The Johns Hopkins University Press; Girard, René 1978. *Things Hidden Since the Foundation of the World*. Stanford: Stanford University Press.

Chapter 4: Jesus as Teacher

1. I acknowledge Rev. Dr. Shawnthea Monroe for helpful insights regarding this parable.

2. Regenstein, Lewis G. 1991. *Replenish the Earth*. New York: Crossroad Publishing; Schwartz, Richard H. 2001. *Judaism and Vegetarianism*. New York: Lantern Books; Linzey, Andrew. 1987. *Christianity and the Rights of Animals*. New York: Crossroad Publishing.

3. Does the Flood contradict the view that God does not want to see creation destroyed? By my reading, God is represented as dismayed by violence and seeing no alternative to destroying most of the earth with a great flood. God's regret about taking such drastic measures was so great that God made a covenant with humanity and all the animals never to flood the earth again.

4. Nuechterlein, Paul J. July 20–21, 1996. Things Hidden Since the Foundation of the World. Last accessed April 5, 2008 from http://girardianlectionary.net/year_a/proper11a_1996_ser.htm.

5. Alison, James. 1996. *Raising Abel: The Recovery of the Eschatological Imagination*. New York: Crossroad Publishing.

6. I am struck by parallels between the parable of the lost sheep and Genesis 18:24–33, in which Abraham asks God whether God would spare Sodom if there were 50 righteous people there. God answered that he would spare the city for the sake of the righteous ones, and Abraham repeatedly asked the question, each time reducing the number of righteous people until he asked a final time, saying, "I will speak again but this once. Suppose ten are found there." God said that he would save the city if there were only ten righteous people. Then God departed, and to my reading the text indicates that Abraham did not dare to ask whether God would save an entire city of sinful people on behalf of a single righteous person. The reason, I think, is that the ancient Hebrews were unprepared to consider that God would abandon his plans for the sake of a single innocent individual. The logic of sacrifice holds that individuals—even innocent individuals—are expendable for the sake of the larger community. Jesus rejected the notion that sacrificing even one individual for the rest of the community accords with God's desires.

7. There are some people who need to eat animals for sustenance, and for them I do not regard meat eating as sinful. However, harming God's creatures unnecessarily, which is the case nearly every time animals are mistreated in the United States, is in my view sinful.

Chapter 5: Jesus as Savior

1. Girard, René. 1972. *Violence and the Sacred.* Baltimore: Johns Hopkins University Press; Girard, René. 1978. *Things Hidden Since the Foundation of the World.* Stanford: Stanford University Press.

2. I prefer to think in gender-neutral terms, because I am convinced that women as well as men must play crucial roles in the reconciliation of God's creation. For example, the New Living Translation uses the term "children" in Romans 8:23, and the King James Version of this verse reads, "And not only they, but ourselves also, which have the first fruits of the Spirit, even we ourselves groan within ourselves, waiting for the adoption, to wit, the redemption of our body."

3. Nuechterlein, Paul J. July 27, 2003. "The God of Abundance." Last accessed April 5, 2008 from http://girardianlectionary.net/year_b/proper12b_2003_ser.htm.

Chapter 6: The Pauline Letters and Christian Faith

1. Nuechterlein, Paul J. 2005. "My Core Convictions: Nonviolence and the Christian Faith." Last accessed April 5, 2008 from http://girardian lectionary.net/core_convictions.htm; Schwager, Raymund. 1987. *Must There Be Scapegoats? Violence and Redemption in the Bible.* New York: Crossroad Publishing, pp. 214–220.

2. Some commentators have argued that the wrath of God was directed at pagans who "exchanged the glory of the immortal God for images resembling mortal man or birds or animals or reptiles" (Romans 1:23). However, to my reading, their offense was failure to "honor" God, and "claiming to be wise, they became fools" (Romans 1:22). They ended up engaging in "dishonorable passions" (Romans 1:26) "and worshiped and served the creature rather than the Creator" (Romans 1:25). In other words, they allowed human desires to distract them from faithfulness to God.

3. Nuechterlein, note 1.

4. *Ibid.*

5. I do think that Paul regarded Jesus as divine.

6. Nuechterlein, note 1.

7. *Ibid.*

8. Alison, James. 2003. *Raising Abel: The Recovery of the Eschatological Imagination.* New York: Crossroad Publishing, p. 108.

Chapter 7: Forgiveness

1. Nuechterlein, Paul J. 2005. "Second Sunday of Easter – Year A." Accessed April 5, 2008 from http://girardianlectionary.net/year_a/easter2a.htm.

2. In Matthew 27:46 and Mark 15:34, Jesus cries, "My God, my God, why hast thou forsaken me?" In ancient times, due to the high cost of paper, authors frequently referred to an entire text by quoting the first line. In this case, the first line of Psalm 22 would direct the reader to the entire psalm, which describes the psalmist's lament followed by a faith-filled proclamation of the psalmist's faith in God.

3. This being said, it is true that Jesus often denounced "hypocrites" (Matthew 6:2, 6:5, 6:16, 22:18, 23:13–29; Mark 7:6–9; Luke 13:15). Sometimes, until people are shocked out of their "comfort zone," they will not see the harm that they do. However, I think one can use powerful words and visual images without accusing anyone of depravity. For further thoughts, see Kaufman, Stephen R. and Nathan Braun. 2004. *Good News for All Creation: Vegetarianism as Christian Stewardship.* Cleveland: Vegetarian Advocates Press, and Adams, Carol J. 2001. *Living Among Meat Eaters.* New York: Three Rivers Press.

4. I acknowledge Julie Shinnick for her contributions to this topic posted at girard.topic@ecunet.org, July 1, 2004 (note #7362).

5. Bailie, Gil (undated). *The Gospel of John* [audiotape series]. Glen Ellen, CA: The Cornerstone Forum.

6. Whatever term people use to describe the divine applies here.

7. Wiesenthal, Simon. 1976. *The Sunflower.* New York: Schocken Books.

8. Storey, Peter (September 10–17, 1997). "A different kind of justice: Truth and reconciliation in South Africa." *The Christian Century* vol. 114, pp. 788–793.

Chapter 8: The Power of Love versus the Power of Satan

1. I thank Norm Phelps for helpful insights here. Among other things, he pointed out that Jesus spoke Aramaic, raising questions about whether Jesus' statements conveyed the same distinctions as those conveyed by the Greek terms *agape* and *philia.* Still, many Jewish men could speak some Greek, which was the regional language of business, politics, and other social relationships.

2. Bailie, Gil. Undated. *The Gospel of John* [audiotape series]. Glen Ellen, CA: The Cornerstone Forum, tape 12.

3. Some have argued that we should include animals among "the least of these." Interestingly, as Rev. Frank Hoffman notes, Jesus might have been

referring to the fish when he asked Peter, "Do you love me more than these?" (John 21:15). When Peter affirmed his love, Jesus replied, "Feed my lambs." We may recall that Peter had been fishing when Jesus called him to become a "fisher of men" (Matthew 4:19; Mark 1:17). According to this interpretation, John 21 describes Jesus asking Peter whether Peter preferred to catch fish or to be a disciple. Jesus had called Peter away from the arguably harmful activity of fishing toward a ministry that involved reconciling all God's creation.

Prior to this exchange with Peter, Jesus demonstrated his divinity by miraculously assisting the disciples in catching fish (John 21:6). Did Jesus not care about the fish? If Jesus killed fish in order to save people, then in essence he was sacrificing one group of innocent individuals, the fish, for others, his disciples. Even if one did not place a high value on fish, they are living, feeling creatures, and their destruction would seem to contradict the notion that Jesus had a nonsacrificial ministry. However, this was a miracle story, and consequently one may envision the miracle consisting not of catching fish but of creating dead fish. See Hoffman, Frank L. "Be Fishers of Men—Not of Fish (John 21:1–17)." Last accessed April 5, 2008 from www.all-creatures.org/discuss/john21.1-17.html.) Indeed, it is just as miraculous to create dead fish as to make a school of fish appear at the side of the boat. Perhaps the reason that the net was not torn (John 21:11) by the huge load was that the fish were not alive and struggling.

Many scholars think that John 21 was a later addition to John's actual Gospel, written in part to prove that Jesus had risen. One reason for this theory is that the final words of John 20, verses 30 and 31 read like concluding comments. Whether or not John 21 describes historical events, the story points toward important truths.

4. I suggest that the reason Jesus said that his blood was poured out for "many," rather than for "all," was that Jesus died—and his "blood was poured out"—for those who sought repentance and forgiveness. For those who refused to repent and receive God's forgiveness, Jesus' blood had no influence on their propensity to victimize innocent individuals.

5. Nevertheless, due to anger or thoughtlessness, we cannot avoid sometimes wounding the ones we love. Though it is often difficult to gauge one's intentions, when the balance of evidently intentional hurtful actions clearly outweighs kind and compassion actions, the relationship is pathological. This balance can be hard to assess, and sometimes counselors can help partners clarify whether a relationship has potential for mutual benefit or whether it is pathological.

6. Although the Greek *huios* regularly translates as "son," the Liddell–Scott–Jones–McKenzie lexicon, which classicists use widely, recognizes "child" as a legitimate translation. I thank Thomas Suits for this observation.

Interestingly, the Revised Standard Version, along with many other translations directed by men, seems to hold a male bias. In the King James Version, Paul sends greetings to two apostles, Andronicus and Junia (Romans 16:7), who had been imprisoned with him. This would indicate that at least one apostle was a woman. *Junia* was a common Greco-Roman name at that time for women, and it appears that the ancient Greek sources favor *Junia* in Romans 16:7. However, the Revised Standard Version and the New International Version use the name *Junias*, with the "ias" ending that indicates a male name. However, this name is not found in ancient, nonbiblical Greek texts. See Ehrman, Bart D. 2006. *Peter, Paul, and Mary Magdalene: The Followers of Jesus in History and Legend.* New York: Oxford University Press; a fuller discussion of evidence favoring Junia is in Epp, Eldon Jay. 2005. *Junia: the First Woman Apostle.* Philadelphia: Augsburg Fortress.

7. Kübler-Ross, Elizabeth. 1991. *On Life After Death.* Berkeley, CA: Celestrial Arts.

8. I am indebted to Rev. Paul J. Nuechterlein for helpful insights in this section, particularly notes on Proper 5 Year B. Last accessed April 5, 2008 from http://girardianlectionary.net/year_b/proper_5b.htm.

9. Girard, René. 2001. *I See Satan Fall Like Lightning.* Maryknoll, NY: Orbis Books.

10. Nuechterlein, Paul J. February 22–23, 1997. "Satan the Accuser and God the Chooser." Last accessed April 5, 2008 from http://girardian lectionary.net/year_b/lent2b_1997_ser.htm.

11. Girard, René. 1986. *The Scapegoat.* Baltimore: The Johns Hopkins University Press, pp. 184–197.

12. Pacifism remains a contentious issue among Christians. Some believe that Christians are called to pacifism regardless of the consequences. Others have argued that refusal to use physical force in defense of vulnerable individuals invites abuse of the innocent. Personally, I think that Christianity calls us to be peacemakers, but I am not a committed pacifist. Although I recognize that the scapegoating process and selfish desires often underlie conflicts, in certain situations it seems reasonable and appropriate to use force, even lethal force, as a last resort to stop greater violence or injustice. If our choices are guided by love and compassion, we will not stray far from righteous paths.

13. Gil Bailie. Undated. *The Gospel of John* [audiotape series], tape 10. Glen Ellen, CA: The Cornerstone Forum; Gospel Communications International. Undated. Commentary: Jesus Speaks of Both His Relation to the Father and His Disciples' Relation to the Father. Last accessed April 5, 2008 from www.biblegateway.com/resources/commentaries/?action=getCommentary Text&cid=4&source=1&seq=i.50.14.2.

14. Heim, S. Mark. 2006. *Saved from Sacrifice: A Theology of the Cross.* Grand Rapids, MI: William B. Eerdmans Publishing, p. 154.

15. Many people expect a life-after-death in heaven in which all our desires will be met. However, mimetic theory tells us that it is not reasonable to view heaven as a place of unlimited resources that satisfies all our desires, because, as acquisitive mimetic creations, much of the reason we derive satisfaction from gaining the objects of desire is that they are scarce. Because so many of our terrestrial desires remain unsatisfied, an everlasting heavenly existence in which all our desires are fulfilled sounds appealing. However, one might anticipate that such a place would eventually become intolerably boring.

16. Dostoyevsky, Fyodor. 1996 [1880]. *The Brothers Karamazov.* New York: The Modern Library, pp. 356–7.

Chapter 9: Healing

1. Scaruffi, Piero. Undated. "Wars and Genocides of the 20th Century." Last accessed April 5, 2008 from www.scaruffi.com/politics/massacre.html.

2. The Greek here reads simply "to them" (presumably the priests), which makes more sense to me. The RSV is distinctive in using "the people" here.

3. Norm Phelps has suggested that Jesus resided in the countryside to avoid the masses of people seeking healing, because too much time and effort dedicated to healing would have interfered with his teaching ministry. See Phelps, Norm. 2002. *The Dominion of Love: Animal Rights According to the Bible.* New York: Lantern Books, pp. 84–85.

4. Animal Place. 2003. *The Emotional World of Farm Animals* [videotape]. Vacaville, CA; Robbins, John. *The Food Revolution.* Berkeley, CA: Conari Press, 2001, pp. 153–164; shnookey [screen name]. "New Piglet. I'm in trouble," and subsequent discussion. Farm Life Forum—Gardenweb. Last accessed April 5, 2008 from forums2.gardenweb.com/forums/load/farmlife/msg111924094139.html?18; Hurley, Blythe, Erica Bernheim, and Angela Mesaros. "Where We Once Were: Stories of Childhood." Last accessed April 5, 2008 from www.keepgoing.org/issue19_demands/where_we_once_were.html; Lush, Tamara. "Cakes, Shakes, and Livestock." *St. Petersburg Times* 2/28/02. Available on the Internet at www.sptimes.com/2002/02/28/TampaBay/Cakes__shakes___and_.shtml; Anonymous. "4-H Boy." Last accessed April 5, 2008 from www.all-creatures.org/poetry/ar-4h.html.

5. Dunayer, Joan. 2001. *Animal Equality: Language and Liberation.* Derwood, MD: Ryce Publishing.

6. Bailie, Gil. Undated. *The Gospel of John* [audiotape series]. Glen Ellen, CA: The Cornerstone Forum.

7. Girard, René. 1986. *The Scapegoat*. Baltimore: The Johns Hopkins University Press, pp. 165–183.

8. Funk, Robert W. and The Jesus Seminar. 1998. *The Acts of Jesus: The Search for the Authentic Deeds of Jesus*. New York: HarperCollins, p. 79.

9. For insightful commentary, see Alison, James. 2001. *Faith beyond Resentment: Fragments Catholic and Gay*. New York: Crossroad, pp. 3–26.

Chapter 10: Peacemaking

1. Heim, S. Mark. 2006. *Saved from Sacrifice: A Theology of the Cross*. Grand Rapids, MI: William B. Eerdmans, p. 102.

2. Akers, Keith. 2000. *The Lost Religion of Jesus: Simple Living and Nonviolence in Early Christianity*. New York: Lantern Books, pp. 120–121.

3. Hays, Richard B. 1996. *The Moral Vision of the New Testament: Community, Cross, New Creation: A Contemporary Introduction to New Testament Ethics*. New York: HarperCollins, pp. 317–346.

4. Wink, Walter. 2003. *Jesus and Nonviolence: A Third Way*. Minneapolis: Fortress Press.

5. Vaclavik, Charles P. 1986. *The Vegetarianism of Jesus Christ*. Three Rivers, CA, Kaweah Publishing; Akers, note 2.

6. Sharp, Gene. *The Politics of Nonviolent Action*. Boston: Extending Horizon Books, 1973.

7. Nuechterlein, Paul J. April 7, 2002. "Dreaming of Peace." Accessed April 5, 2008 from http://girardianlectionary.net/year_a/easter2a_2002_ser.htm.

Chapter 11: Christian Faith and Prophetic Witness

1. I acknowledge helpful insights here from Rev. James Antal, former pastor of Plymouth Church of Shaker Heights, UCC and currently Conference Minister and President, Massachusetts Conference, United Church of Christ.

2. Nuechterlein, Paul J. August 14 & 18, 2002. "Faith is Not Being Scandalized." Last accessed April 5, 2008 from www.girardian lectionary.net/year_a/proper15a_2002_ser.htm.

3. Kushner, Harold S. 1983. *When Bad Things Happen to Good People*. New York: Avon.

4. Nuechterlein, Paul J. "Proper 22—Year B: General Commentary: Equality and Mimetic Desire." Last accessed April 5, 2008 from http://girardianlectionary.net/year_b/proper22b.htm.

5. Loy, David. *Nonduality: A Study in Comparative Philosophy.* 1988. New Jersey; Humanities Press; Shlain, Leonard. 1998. *The Alphabet versus the Goddess: The Conflict between Word and Image.* New York. Viking Penguin; Grandin, Temple and Catherine Johnson. 2005. *Animals in Translation: Using the Mysteries of Autism to Decode Animal Behavior.* New York: Scribner.

Grandin has discussed how, as a consequence of her autism, she thinks with images rather than language. She has maintained that animals similarly think with images, and she has gained a reputation as an expert in understanding and treating animal behavior problems and at devising ways to reduce animal stress on farms and at slaughterhouses.

6. Though prophecy carries substantial risks, not all prophets have been killed. Jesus was likely being hyperbolic here, but his basic point is true.

7. Spiegel, Marjorie. 1988. *The Dreaded Comparison: Human and Animal Slavery.* New York: Mirror Books.

8. Patterson, Charles. 2002. *Eternal Treblinka: Our Treatment of Animals and the Holocaust.* New York: Lantern Books; Sztybel, David. 2006. "Can the Treatment of Animals Be Compared to the Holocaust?" *Ethics and the Environment* vol. 11, pp. 97–132.

9. Bailie, Gil. 1997. *Violence Unveiled.* New York: Crossroad Publishing, pp. 167–184.

10. Tannehill, Robert C. 1986. *The Narrative Unity of Luke–Acts.* Philadelphia: Fortress Press, p. 37.

11. Coffin, William Sloane. 2004. *Credo.* Louisville, KY: Westminster John Knox Press, p. 49.

12. On what grounds might one claim certainty about the nature of God? One can establish attributes of God as fundamental statements of faith, but this does not constitute evidence for one's claims. Indeed, many people accept a particular set of beliefs about God "on faith," but this seems to merely reflect prejudice for the tenets of one's culture. Someone raised in a different culture would be equally justified in adhering to a vastly different set of beliefs about God. Many people relate personal experiences of divine presence, but our hopes, fears, and prejudices can color our experiences and how we interpret them. Furthermore, the brain can influence the mind, and people with temporal lobe epilepsy sometimes relate experiences of divine presence that often result in persistent, firm convictions about God, the afterlife, and other religious issues. (See Dewhurst, Kenneth and A.W. Beard. 1970. "Sudden religious conversions in temporal lobe epilepsy." *British Journal of Psychiatry* vol. 117, pp. 497–507.) Another approach is to derive conclusions about God's nature by carefully observing God's world and its inhabitants. This approach might lead to

reasonable conclusions about God's nature, but empirical observations cannot provide certainty. Our senses and our ability to interpret data are imperfect, and there is always the possibility that new evidence could contradict our theories.

Chapter 12: The New Testament and Sacrifice

1. I thank Cindi McAndrew, Gary Neitzke, and Paul Dobberstein for helpful commentary on Hebrews 10:13, which they provided at the Girardian discussion list girard.topic@ecunet.org.

2. Hardin, Michael. "Sacrificial Language in Hebrews: Reappraising René Girard," in Swartley, William M., ed. 2000. *Violence Renounced: René Girard, Biblical Studies, and Peacemaking.* Telford, PA: Pandora Press, pp. 103–119.

3. *Ibid*, p. 114.

4. *Ibid*, p. 115.

5. Howard-Brook, Wes and Anthony Gwyther 2000. *Unveiling Empire: Reading Revelation Then and Now.* New York: Maryknoll, pp. 117–118.

6. *Ibid*, pp. xxvii-xxviii, 148–149.

7. Bailie, Gil. 1997. *Violence Unveiled: Humanity at the Crossroads.* New York: Crossroad Publishing, p. 15.

8 Howard-Brook and Gwyther, note 5, pp. 117–118.

9. Webb, Eugene. 2005. "René Girard and the symbolism of religious Sacrifice." *Anthropoetics* 11 (1). Available on the Internet at www.anthropoetics.ucla.edu/ap1101/webb.htm; Sensing, Donald. March 11, 2004. "A short history of Original Sin." Last accessed April 5, 2008 from thereligiousnews.blogspot.com/2004_03_07_the religiousnews_archive.html.

10. I would like to emphasize that either way of regarding the Bible, literally or allegorically, encourages love and compassion in all our dealings with humans and animals.

11. Becoming human involves developing human self-consciousness, as discussed in Chapter 2. Human self-consciousness plays a crucial role in humanity's distinctive (but not necessarily exclusive) fear of death, need for self-esteem, and acquisitive mimetic desire.

12. Kirk, Peter. "Augustine's Mistake about Original Sin." Last accessed April 5, 2008 from qaya.org/blog/?p=246; Andrew [computer screen name]. "Evolution of Doctrine: Original Sin." Last accessed April 5, 2008 from theogeek.blogspot.com/2007/09/evolution-of-doctrine-original-sin.html.

13. Becker, Ernest. 1975. *Escape from Evil.* New York: The Free Press.

14. Akers, Keith. 2000. *The Lost Religion of Jesus: Simple Living and Nonviolence in Early Christianity.* New York: Lantern Books; Vaclavik, Charles P. 1986. *The Vegetarianism of Jesus Christ.* Three Rivers, CA, Kaweah Publishing.

15. Roman law had exempted the Jews from worshipping the emperor, because the Jews had made it clear that they would rather die than betray their faith. When the Jewish Christian movement broke away from the synagogue, its members no longer enjoyed the protection of the "Jewish exception."

16. Pagels, Elaine. 1988. *Adam, Eve, and the Serpent.* New York: Random House, p. 73; Webb, Eugene. 1992. "Augustine's New Trinity: The Anxious Circle of Metaphor," in Williams, Michael A., Collett Cox, and Martin S. Jaffee. *Religious Innovation: Essays in Interpretation of Religious Change.* Berlin: Mouton de Gruyter, 1992, pp. 191–214.

17. Pagels, note 16, pp. 98–99.

18. *Ibid*, p. 73.

19. Hall, Douglas John. 2003. *The Cross in Our Context: Jesus and the Suffering World.* Minneapolis: Fortress Press, 2003, pp. 171–172.

20. Weaver, J. Denny. 2001. "Violence in Christian Theology." *Cross Currents* July 2001. Available on the Internet at www.crosscurrents.org/weaver0701.htm; Weaver, J. Denny. 2001. *The Nonviolent Atonement.* Grand Rapids, MI: W.B. Eerdmans; Schwager, Raymond. 1999. *Jesus in the Drama of Salvation.* New York, Crossroad Publishing.

21. Proponents of substitutionary atonement theology often point to Romans 3:25 for biblical support. To understand this passage, it is helpful to consider Romans 3:21–26, which reads as follows:

[21]But now the righteousness of God has been manifested apart from the law, although the law and the prophets bear witness to it, [22]the righteousness of God through faith in Jesus Christ for all who believe. For there is no distinction; [23]since all have sinned and fall short of the glory of God, [24]they are justified by his grace as a gift, through the redemption which is in Jesus Christ, [25]whom God put forward as an expiation by this blood, to be received by faith. This was to show God's righteousness, because in his divine forbearance he has passed over former sins; [26]it was to prove at the present time that he himself is righteous and that he justifies him who has faith in Jesus.

Let us first consider the notion of "expiation" of sin. Through the lens of substitutionary atonement theology, one might regard punishment of the sinner or a substitute victim as the only way to balance the scales of justice, expiate sin, and restore justification with God. (See Bartlett, Anthony W. 2001. *Cross Purposes: The Violent Grammar of Christian Atonement*. Harrisburg, PA: Trinity Press.) However, substitutionary atonement theology developed over many centuries after Jesus' death; does this theology accurately reflect what Paul tried to communicate in his letter to the Romans? Another view, which I think reflects better the life and ministry of Jesus, is that expiation of sin involves removing sinfulness from one's own soul and from one's community.

In support of this notion of expiation, recall that Romans 3:22 and 3:26 can very reasonably be translated as "faith of Christ" (see Chapter 6). Romans 3:21–22 indicates that having the faith of Christ, rather than abiding by the law and its sacrificial practices, is the way to manifest the righteousness of God. Everyone, including believers, continue to sin (verse 23), and nobody merits justification by virtue of works. Justification is a gift, and Jesus Christ was the vehicle for delivering that gift (verse 24).

If we regard Jesus' death as a self-sacrifice (see Chapter 3), verse 25 takes on a very different meaning from that suggested by substitutionary atonement theology. God called Jesus to show people how to build communities grounded on love rather than on scapegoating and victimization. Jesus chose to accept this destiny, even to the point of death. For Jesus to be faithful to his divine mission, he had to give his blood—which the ancients regarded as the source of life. For Christians, those who have the faith of Christ strive to reflect Jesus' faithfulness and service to God.

How could Jesus death expiate sin? Paul related, "In his [God's] divine forbearance he has passed over former sins," indicating that God offers forgiveness for sins without requiring sacrifices. Proponents of substitutionary atonement theology tend to argue that the reason that sacrifices are now unnecessary is that the sacrifice of Jesus ended the need for sacrifices to expiate sin. However, Jesus forgave sins on behalf of God (Matthew 9:6; Mark 2:5; Luke 5:20; John 8:11) without sacrificing or harming anyone.

Jesus taught followers how to resist the human tendency to participate in the sins of scapegoating and victimization, and he struggled against the injustices of his day. Likewise, Christians are called to defend those who are vulnerable or victims of injustice, even to the point of self-sacrifice. Sin is not just an individual problem; it is a *communal* problem. Human sinfulness often reflects our experiences, particularly in childhood, that strongly influence whether or not we will make kind, compassionate choices. Further, we tend to fall into sin when we become gripped by acquisitive mimetic desires, which require other people and occur in the context of community. Finally, we contribute to others' sinfulness when we are insensitive, callous, or rapacious. While we cannot free ourselves or our communities from sin, Christians can expiate their own past, present and future sins, as well as relieve the sins of their communities, by striving to reflect the faith of Christ. Verse 26 concludes that God is righteous and God justifies those who have the faith of Christ—a view that works against the scapegoating process.

This analysis of Romans 3:25 similarly applies to "expiation" of sin in 1 John 2:2 and 4:10. Though it is impractical to attempt to address all biblical verses highlighted by proponents of substitutionary atonement theory, I have discussed elsewhere in this book many of the passages often cited by defenders of this theory, such as Genesis 22:13, Isaiah 53:4-5, Romans 6:23, and 2 Corinthians 5:21. Another frequently cited verse is Romans 4:25, which reads, "[Jesus] was put to death for our trespasses and raised for our justification." A Girardian reading that does not involve substitutionary atonement sees Jesus' death as a consequence of our trespasses— notably our tendency to participate in the scapegoating process. "Raised for our justification" refers to how the Resurrection was a crucial event for Christians becoming right with God (see Chapter 3).

22. If we are to regard Jesus' death as a sacrifice in the context of the Jewish sacrificial cult, then it is important to consider the purpose of the atonement sacrifices for the Hebrews. The ancient Hebrews' did not regard sacrifices for sin as designed to appease God's wrath—a notion that is implicit in substitutionary atonement theology. James D.G. Dunn has noted that the act of sacrifice denoted by the Hebrew verb *kipper* aimed to eradicate the sin or the sinner's propensity to sin. (*Kipper* relates to *Yom Kippur*, the holiest day of the Hebrew calendar and means "Day of Atonement.") Dunn wrote,

> But in Hebrew usage God is never the object of the key verb (*kipper*). Properly speaking, in the Israelite cult, God is never "propitiated" or "appeased." The objective of the atoning act is rather the removal of sin—that is, either by purifying the person or object, or by wiping out the sin. Atonement is characteristically made "for" a person or "for sin." . . . Of course, the atoning act thus removes the sin which provoked God's wrath, but it does so by acting on the sin rather than on God.

Dunn, James D.G. 1998. *The Theology of Paul the Apostle*. Grand Rapids, MI: Wm. B. Eerdmans, p. 214.

23. Weaver has also noted that scientists no longer regard the essence of life to reside in the blood, but that the ancient Hebrew understanding does have physiologic foundation insofar as an animal without blood dies.

24. Weaver, *The Nonviolent Atonement*, p. 59.

25. *Ibid.*

26. Weaver, "Violence and Christian Theology."

27. *Ibid.*

28. The symbolism in Revelation shows that the writer equated the Roman Empire with the forces of evil. The seven-headed dragon (Revelation 12:3) relates to the Seven Hills of Rome as well as a sequence of seven emperors. See, for example, Howard-Brook and Gwyther, note 5, pp. 117–118.

29. Weaver, "Violence and Christian Theology."

Chapter 13: Contemporary Issues

1. Henshaw, Stanley K., Susheela Singh, and Taylor Haas. 1995. "The Incidence of Abortion Worldwide." *Family Planning Perspectives* 25 (supplement). Last accessed April 5, 2008 from www.guttmacher.org/pubs/journals/25s3099.html.

2. Salter, Cynthia, Heidi Bart Johnson, Nicolene Hengen, and Ward Rinehart, eds. 1997. "The Extent of Unsafe Abortion, in Population Information Program." Center for Communication Programs, The Johns Hopkins School of Public Health. *Population Reports: Care for Postabortion Complications: Saving Women's Lives.* Volume 25, no. 1. Last accessed April 5, 2008 from www.infoforhealth.org/pr/l10/l10chap1_2.shtml.

3. Hyland, J.R. 1995. *Sexism is a Sin: The Biblical Basis of Female Equality.* Sarasota, FL, Viatoris Ministries.

4. Keen, Sam. 1991. *Fire in the Belly: On Being a Man.* New York: Bantam Books; Beers, William. 1992. *Women and Sacrifice: Male Narcissism and the Psychology of Religion.* Detroit: Wayne State University Press.

5. Shapiro, Kenneth J. 1990. "Animal Rights versus Humanism: The Charge of Speciesism." *Journal of Humanistic Psychology* vol. 30 (no. 2), pp. 9–37.

6. Bailie, Gil. Undated. *Reflections on the Gospel of John* [audiotape series]. Glen Ellen, CA: The Cornerstone Forum, tape 4.

7. *Ibid.*

8. General Council of the Assemblies of God. Homosexual Conduct. Last accessed April 5, 2008 from www.ag.org/top/Beliefs/relations_11_homosexual.cfm; Feeney, Pastor Jim. "Biblical Arguments Against Gay Marriage." Last accessed April 5, 2008 from www.jimfeeney.org/againstgaymarriage.html.

9. Cannon, Justin R. "The Bible, Christianity, & Homosexuality." Accessed

April 5, 2008 from www.truthsetsfree.net/bible.htm; "Does the Bible Condemn Homosexuality? An Interview with Dr Reverend Cheri DiNovo." *The Turning Magazine*. Accessed April 5, 2008 from www.theturning.org/folder/samesex.html; Helminiak, Daniel A. 2002. *What Does the Bible Really Say about Homosexuality*. New Mexico, Alamo Square Press. Helminiak argues that Genesis 19:1–11 condemns the intent of the Sodomites to rape the visitors. The evil intent of the wicked people of Sodom to gang rape the visitors contrasts with Abraham's abundant hospitality toward these angels prior to their visit to Sodom (Genesis 18:2–8). Ezekiel relates the reason the Sodomites were condemned: "Behold, this was the guilt of your sister Sodom: she and her daughters had pride, surfeit of food, and prosperous ease, but did not aid the poor and needy." (16:49). Leviticus 18:22 and 20:13 prohibit and condemn male homosexual acts among the ancient Hebrews, which, like the kosher dietary laws, was one of the ways the ancient Hebrews separated themselves from other cultures. This ancient Hebrew prohibition should not be understood as a general condemnation of homosexuality. "Unnatural" in Romans 1:26 is better translated as "atypical" [but many translators would consider *para physia* as closer to "contrary to nature"]. The evident condemnation of homosexuality in Romans 1:24–27 is more reasonably regarded as an observation that homosexuality is degrading, shameful, and dishonorable, but it is *not* immoral or sinful, unlike the sinful activities described in Romans 1:29–31. The real "error" (Romans 1:27) was idolatry (Romans 1:21–23), because idolatry leads to all kinds of degrading activities and sin. Finally, while it is unclear whether 1 Corinthians 6:9 and 1 Timothy 1:10 refer to homosexuality specifically or sexuality in general, these passages condemn exploitative, abusive, or wanton sex—not all homosexuality or heterosexuality.

10. Republican Party of Texas. "What We Believe: 2006 State Republic Party Platform." Last accessed April 5, 2008 from texasgop.org/site/DocServer/2006_Plat_with_TOC_2.pdf?docID=2022.

11. Johnson, Britton W. "A Proposal to Use Girardian Anthropology to Analyze and Resolve the Present Challenge to the 'Peace, Unity and Purity of the Church.' " Last accessed April 5, 2008 from brittondanna.wordpress.com/files/2008/03/peaceunitypurityreflection.pdf.

12. *Ibid.*

13. Malthus, Robert. 1999 [1798]. *An Essay on the Principle of Population*. New York: Oxford University Press.

14. I thank Rev. Linda McDaniel for helpful insights in this paragraph.

15. Bailie, Gil. Undated. *The Gospel of John* [audiotape series]. Glen Ellen, CA: The Cornerstone Forum, tape 9.

16. There have been significant military adventures in the past few decades that have put Americans in harm's way overseas, but foreign powers have not threatened to conquer the United States.

17. Leifer, Ron. 1997. *The Happiness Project.* Ithaca: Snow Lion Publications. I do not think one should equate liberalism with the Democratic Party nor conservatism with the Republican Party. The policies of both political parties embrace both ideologies to varying degrees.

18. Johnson, note 11.

19. *Ibid.*

20. Cooney, Robert P.J. Jr. 2005. *Winning the Vote: The Triumph of the American Women's Suffrage Movement.* Santa Cruz, CA: American Graphic.

21. Singer, Peter. 1991/1975. *Animal Liberation.* New York: Avon Books.

22. Regan, Tom. 1983. *The Case for Animal Rights.* Berkeley, CA: University of California Press.

23. Scully, Matthew. 2002. *Dominion: The Power of Man, the Suffering of Animals, and the Call to Mercy.* New York: St. Martin's Press.

24. Many animal-use industries have claimed to be concerned about animal welfare, but it has been my impression that this has generally been false posturing.

25. Dunayer, Joan. 2001. *Animal Equality: Language and Liberation.* Derwood, MD: Ryce Publishing.

26. Youths involved in 4-H or similar programs typically name the animals for whom they must care until the animal is sold for slaughter. They often find betraying the animal's friendship and trust heartbreaking. Evidently in an effort to desensitize the youths, adults often encourage them to give names that relate to the animal's destiny, such as "Sausage Patty" (the title of a nice children's book by Diane Allevato, published by Animal Place, 1998).

27. Similarly, animal experimenters typically identify individual animals by their species name and assigned number, such as "monkey 212." Animal experimenters and other people engaged in activities that harm animals generally refer to individual animals with the impersonal pronoun "it" rather than "he" or "she." See Dunayer, Joan, note 25; Phillips, Mary T. 1991. *Constructing Laboratory Animals: An Ethnographic Study in the Sociology of Science* [dissertation in the Department of Sociology of New York University].

28. Mangels, Ann Reed, Virginia Messina, and Vesanto Melina. 2003. "Position of the American Dietetic Association and Dietitians of Canada:

Vegetarian Diets." *Journal of the American Dietetic Association* vol. 103, pp. 748–765; Barnard, Neal. 1993. *Food for Life: How the New Four Food Groups Can Save Your Life.* New York: Three Rivers Press.

29. Worldwatch Institute. 2004. "Meat: Now, It's Not Personal!" *World Watch* July/August, pp. 12–20; Sapp, Amy. "Production and Consumption of Meat: Implications for the Global Environment and Human Health," Last accessed April 5, 2008 from www.med.harvard.edu/chge/course/ papers/sapp.pdf; The University of Chicago News Office. "Study: vegan diets healthier for planet, people than meat diets." Last accessed April 5, 2008 from www-news.uchicago.edu/releases/06/060413.diet.shtml; Sierra Club. Clean Water & Factory Farms. Last accessed April 5, 2008 from www.sierraclub.org/factoryfarms/.

30. Gregor, Michael. 2006. *Bird Flu: A Virus of Our Own Hatching.* New York: Lantern Books, 2006; Lyman, Howard F. 1998. *Mad Cowboy.* New York: Scribner.

31. Kalechofsky, Roberta. *The Poet-Physician and the Healer-Killer: The Emergence of a Medical Technocracy* [in progress]; Lederer, Susan E. 1995. *Subjected to Science: Human Experimentation in America before the Second World War.* Baltimore: Johns Hopkins University Press; Advisory Committee on Human Radiation Experiments. 1995. *The Human Radiation Experiments.* New York: Oxford University Press.

32. More people die of preventable infectious diseases than direct human violence, but many of those who die of preventable infectious disease succumb because human violence has displaced them from sources of food and safe water.

33. There are injustices associated with the way Americans eat, because the inefficiencies of animal agriculture make it impossible for the world to consume the typical American diet. Further, contemporary agriculture is not sustainable. All farming requires nitrogen fertilizer. Traditional farms produce a range of plant and animal foods, and animal manure provides the nitrogen for plants. However, manure produced in modern concentrated, animal-feeding operations is discarded as waste and results in environmental damage. Most intensive farmers use large quantities of industrial nitrogen fertilizer, which runs off into streams and kills aquatic life. (See Pollan, Michael. 2006. *The Omnivore's Dilemma: A Natural History of Four Meals.* New York: Penguin Press.) Traditional farming, unlike modern intensive farming, is environmentally sustainable, but it is more labor intensive and provides lower short-term yields. In order to feed the people of today and tomorrow, we will need both population control and a move toward plant-based diets in the West.

34. Marcus, Erik. 2005. *Meat Market: Animals, Ethics, and Money*. Ithica, NY: Brio Press, pp. 187–188.

35. Robbins, John. 2001. *The Food Revolution*. Berkeley, CA: Conari Press.

36. Steinfeld, Henning, Pierre Gerber, Tom Wassenaar, Vincent Castel, Mauricio Rosales, and Cees de Haan. 2006. *Livestock's Long Shadow: Environmental Issues and Options*. Rome: Food and Agriculture Organization of the United Nations. Available on the Internet at www.virtualcentre.org/en/library/key_pub/longshad/A0701E00.htm.

37. Nearly all religions have tenets that point toward love, peace, and nonviolence. A person does not necessarily have to be Christian to abide by a faith analogous to the faith of Christ. A society grounded on the principles Jesus taught and exemplified does not need to have its members regard themselves as Christian. See www.vegsource.com/biospirituality/main.html, last accessed April 5, 2008.

38. I acknowledge Stephen H. Webb for insights here.

39. Sharp, Gene, 1973. *The Politics of Nonviolent Action* [three-part monograph]. Boston: Extending Horizons Books. Branch, Taylor. *Parting the Waters: America during the King Years, 1954–63*. New York: Simon and Shuster, 1989.

40. Benjamin, Rich and Jamie Carmichael. "What King really dreamed." *The Boston Globe* Jan.15, 2006. Available on the Internet at www.boston.com/news/globe/editorial_opinion/oped/articles/2006/01/15/what_king_really_dreamed/.

41. Bill Moyers. "Politics and Economy." Last accessed April 5, 2008 from www.pbs.org/now/politics/poverty_pop/index.html.

42. Moyers, Bill. June 23, 2007. "Drive out the Money Changers" [Presentation at the United Church of Christ General Synod]. Last accessed April 5, 2008 from www.ucc.org/news/significant-speeches/moyers-challenges-ucc-drive.html.

43. Benjamin and Carmichael, note 39.

44. Gibbon, Edward (ed. by Bury, J. B.) 2004 [1776–1788]. *The History of the Decline and Fall of the Roman Empire*. Holicong, PA: Wildside Press; Diamond, Jared M. 2005. *Collapse: How Societies Choose to Fail or Succeed*. New York: Viking.

45. Klerman, G. L. and M. M. Weissman. 1989. "Increasing rates of depression." *Journal of the American Medical Association* vol. 261, pp. 2229–2235; Greenberg, Paul E., Ronald C. Kessler, Howard G. Birnbaum, et al. 2003. "The Economic Burden of Depression in the United States: How Did It Change Between 1990–2000?" *Journal of Clinical Psychiatry*

vol. 64, pp. 1465–1475.

46. William Sloane Coffin wrote,

> That God is against the status quo is one of the hardest things to believe if you are Christian who happens to profit by the status quo. In fact, most of us don't really believe it, not in our heart of hearts. We comfort ourselves with the thought that because our intentions are good (nobody gets up in the morning and says, "Whom can I oppress today?"), we do not have to examine the consequences of our actions. As a matter of fact, many of us are even eager to respond to injustice, as long as we can do so without having to confront the causes of it.

Coffin, William Sloane. 2004. *Credo*. Louisville, KY: Westminster John Knox Press, p. 64.

47. Becker, Ernest. 1975. *Escape from Evil*. New York: The Free Press.

Valuable Resources

The following resources have aided in my study of violence. The list is by no means exhaustive, and indeed a wide range of fictional and non-fictional literature has helped me think about this perennial problem.

Alford, C. Fred. 1997. *What Evil Means to Us*. Ithica, NY: Cornell University Press.

Alison, James. 2003. *Raising Abel: The Recovery of the Eschatological Imagination*. New York: Crossroad Publishing.

— 1998. *The Joy of Being Wrong: Original Sin Through Easter Eyes*. New York: Crossroad Publishing.

Ayers, Ed. 1999. *God's Last Offer: Negotiating for a Sustainable Future*. New York: Four Walls Eight Windows.

Bailie, Gil. 1997. *Violence Unveiled: Humanity at the Crossroads*. New York: Crossroad Publishing.

— The Cornerstone Forum [Web site], www.cornerstoneforum.org/.

Bartlett, Anthony W. 2001. *Cross Purposes: The Violent Grammar of Christian Atonement*. Harrisburg, PA: Trinity Press.

Becker, Ernest 1971 [1962]. *The Birth and Death of Meaning: An Interdisci-plinary Perspective on the Problem of Man*. New York: The Free Press.

— 1968. *The Structure of Evil*. New York: George Braziller.

— 1971. *Birth and Death of Meaning, 2nd Ed*. New York: The Free Press.

— 1973. *The Denial of Death*. New York: The Free Press.

— 1975. *Escape from Evil*. New York: The Free Press.

Branch, Taylor. *Parting the Waters: America during the King Years, 1954–63*. New York: Simon and Shuster, 1989.

Brown, Norman O. 1959. *Life Against Death: The Psychoanalytical Meaning of History*. Hanover, NH: Wesleyan University Press.

Diamond, Stephen A. 1996. *Anger, Madness, and the Daimonic: The Psychological Genesis of Violence, Evil, and Creativity*. Albany, NY: State University of New York Press.

Esselstyn, Caldwell. 2007. *Prevent and Reverse Heart Disease*. New York: Avery.

Fleming, Chris. 2004. *René Girard: Violence and Mimesis*. Malden, MA: Polity Press.

Fox, Matthew. 1988. *The Coming of the Cosmic Christ*. San Francisco: HarperSanFrancisco.

Fromm, Erich. 1973. *The Anatomy of Human Destructiveness*. New York: Holt, Rinehart, & Winston.

Gandhi, Mohandas K. 1927. *An Autobiography*. Ahmedabad, India: Navajivan Trust.

Gieryn, Thomas F. 1994. "Objectivity for these times." *Perspectives on Science* vol. 2, pp. 324–349.

Gilligan, Carol. 1982. *In a Different Voice: Psychological Theory and Women's Development*. Cambride, MA: Harvard University Press.

Girard, René. 1977. *Violence and the Sacred*. Baltimore: Johns Hopkins University Press.

— 1978. *Things Hidden Since the Foundation of the World*. Stanford: Stanford University Press.

— 2001. *I See Satan Fall Like Lightning*. Maryknoll, NY: Orbis Books.

— 1996. (ed. by James G. Williams) *The Girard Reader*. New York: Crossroad Publishing.

— 1986. *The Scapegoat*. Baltimore: The Johns Hopkins University Press.

— 1965. *Deceit, Desire, and the Novel*. Baltimore: The Johns Hopkins University Press.

Hanh, Thich Nhat. 1995. *Living Buddha, Living Christ*. New York: Riverhead Books.

Harden, Michael. *Preaching Peace* [Web site], www.preachingpeace.org.

Heim, S. Mark. 2006. *Saved from Sacrifice: A Theology of the Cross*. Grand Rapids, MI; William B. Eerdmans.

Hammerton-Kelly, Robert G. 1994. *The Gospel and the Sacred: Poetics of Violence in Mark*. Minneapolis, MN: Fortress Press.

Hyland, J.R. 1995. *Sexism is a Sin: The Biblical Basis of Female Equality*. Sarasota, FL, Viatoris Ministries.

Keen, Sam. 1994. *Hymns to an Unknown God: Awakening the Spirit in Everyday Life*. New York, NY: Bantam Books.

— 1992. *Fire in the Belly: On Being a Man*. New York: Bantam Books.

Leifer, Ron. 1997. *The Happiness Project: Transforming the Three Poisons That Cause the Suffering We Inflict on Ourselves and Others*. Ithaca, NY: Snow Lion Publications.

— 1999. "Buddist Conceptualization and Treatment of Anger." *Psychotherapy in Practice* 55 (3):339-351.

— January 2004. *The Roots of War* [presentation at Symposium on the Psychological Interpretation of War in New York City].

Leifer, Ron. 2008. *Vinegar into Honey: Seven Steps to Understanding and Transforming Anger, Aggression, and Violence.* Ithica, NY: Snow Lion.

Liechty, Daniel. Undated. *Towards Concurrence and Collaboration: A Beckerian Looks at Salient Aspects of le Systeme Girard* [presentation at Syracuse University].

Liechty, Daniel (ed.), *The Ernest Becker Reader.* Seattle: The Ernest Becker Foundation, 2005.

Lifton, Robert Jay. 1986. *The Nazi Doctors: Medical Killing and the Psychology of Genocide.* New York: Basic Books.

Linzey, Andrew. 1994. *Animal Theology.* Urbana, IL: University of Illinois Press.

Loy, David R. 2003. *The Great Awakening: A Buddhist Social Theory.* Somerville, MA: Wisdom Publications.

— 2002. *A Buddhist History of the West: Studies in Lack.* Albany, NY: State University of New York Press.

— 1992. "Trying to Become Real: A Buddhist Critique of Some Secular Heresies." *International Philosophical Quarterly* 32 (4):403-425.

— undated. "The Spiritual Roots of Modernity: Buddhist Reflections on the Idolatry of the Nation-State, Corporate Capitalism and Mechanistic Science." Last accessed April 5, 2008 from www.bpf.org/tsangha/loy-roots.html.

Marr, Andrew. "Seeking Peace: A Benedictine Web Site with Articles on Prayer and Spirituality." Last accessed April 5, 2008 from www.andrewmarr.homestead.com/index~ns4.html.

May, Rollo. 1969. *Love and Will.* New York: Bantam Doubleday Dell Publishing Group.

— 1981. *Freedom and Destiny.* New York: Bantam Doubleday Dell Publishing Group.

— 1972. *Power and Innocence: A Search for the Sources of Violence.* New York: W.W. Norton & Co.

— 1991. *The Cry for Myth.* New York: Bantam Doubleday Dell Publishing Group.

Midgley, Mary. 1995. *Beast and Man: The Roots of Human Nature.* London: Routledge.

— 1986. *Wickedness, a Philosophical Essay.* London: Ark Paperbacks.

— 1994. *The Ethical Primate: Humans, Freedom and Morality*. London: Routledge.

Miles, Jack. 1995. *God: A Biography*. New York: Alfred A. Knopf.

Nuechterlein, Paul J. Girardian *Reflections on the Lectionary*. http://girardianlectionary.net/.

— "My Core Convictions: Nonviolence and the Christian Faith." Last accessed April 5, 2008, http://girardianlectionary.net/core_convictions.htm.

— "René Girard: The Anthropology of the Cross as Alternative to Post-Modern Literary Criticism." Last accessed April 5, 2008 from http://girardianlectionary.net/girard_postmodern_literary_criticism.htm.

Phelps, Norm. 2002. *The Dominion of Love: Animal Rights According to the Bible*. New York: Lantern Books.

Pollan, Michael. 2006. *The Omnivore's Dilemma: A Natural History of Four Meals*. New York: Penguin Press.

Preece, Rod. 1999. *Animals and Nature: Cultural Myths, Cultural Realities*. Vancouver, UBC Press.

Regan, Tom, ed. 1986. *Animal Sacrifices: Religious Perspectives on the Use of Animals in Science*. Philadelphia: Temple University Press.

Rhodes, Richard. 1999. *Why They Kill: The Discoveries of a Maverick Criminologist*. New York: Alfred A. Knopf.

Schwager, Raymund. 2000 [1978]. *Must There Be Scapegoats? Violence and Redemption in the Bible*. New York: Crossroad Publishing.

— 1999. *Jesus in the Drama of Salvation*. New York: Crossroad Publishing.

Schwartz, Richard. 2002. *Judaism and Global Survival*. New York: Lantern Books.

Scully, Matthew. 2002. *Dominion: The Power of Man, the Suffering of Animals, and the Call to Mercy*. New York: St. Martin's Press.

Sharp, Gene, 1973. *The Politics of Nonviolent Action* [three-part monograph]. Boston: Extending Horizons Books.

Swartley, William M. 2000. *Violence Renounced: René Girard, Biblical Studies, and Peacemaking*. Telford, PA: Pandora Press.

Sztybel, David. 2007. *Acts of God*. Published on the Internet at http://sztybel.tripod.com/ag.htm.

Wallace, Mark I., and Theophus H. Smith, eds. 1994. *Curing Violence*. Sonoma, CA: Polebridge Press.

Webb, Stephen H. 1998. *On God and Dogs: A Christian Theology of*

Compassion for Animals. New York: Oxford University Press.

Webb, Stephen H. 2001. *Good Eating*. Grand Rapids, MI: Brazos Press.

Williams, James G., ed., 1996. *The Girard Reader*. New York: Crossroad Publishing.

– 1991. *The Bible, Violence, and the Sacred: Liberation from the Myth of Sacred Violence*. Valley Forge, PA: Trinity Press International.

Wright, Richard. 1994. *The Moral Animal: The New Science of Evolutionary Psychology*. New York: Pantheon Books.

Young, Richard Alan. 1999. *Is God a Vegetarian? Christianity, Vegetarianism, and Animal Rights*. Chicago and La Salle, IL: Open Court.

Index

Scripture Index

About the Author

Stephen R. Kaufman, M.D. is an ophthalmologist specializing in retinal diseases and is an Assistant Professor of Ophthalmology at Case School of Medicine. He is chair of the Christian Vegetarian Association, president of Vegetarian Advocates (a Cleveland-based group), and cochair of the Medical Research Modernization Committee. He has written and lectured on Ernest Becker and René Girard, scientific shortcomings of animal experimentation, diet and health, and vegetarianism and Christianity. He is a lay member of the United Church of Christ, a Protestant denomination. He is married, lives in Shaker Heights, Ohio, and has two grown sons.

Also from Vegetarian Advocates Press

Good News for All Creation: Vegetarianism as Christian Stewardship by Stephen R. Kaufman and Nathan Braun, 2004, 123 pp.

"If you need a biblical mandate for changing your diet, this book will meet that need. It is important to read for your own good, for the good of the world, and for God's sake."
   —Tony Campolo, Professor of Sociology, Eastern College,
      St. Davids, PA

"*Good New for All Creation* is an effective and powerful testimony that becoming vegetarian strengthens, rather than weakens, one's Christian beliefs and one's personal witness of Christ's compassion."
   —Keith Akers, author of *The Lost Religion of Jesus*

Every Creature a Word of God: Compassion for Animals as Christian Spirituality by Annika Spalde and Pelle Strindlund, 2008, 162 pp.

"Gracefully combining balanced scholarship with personal witness, animal activists Annika Spalde and Pelle Strindlund have written a book that will enable Christians of all denominations to rediscover the powerful tradition of creaturely compassion that runs throughout their religious history."
   —Reverend Gary Kowalski, author of *The Souls of Animals* and
      The Bible According to Noah: Theology as if Animals Mattered

"Many books about animals, diet, and Christianity have been written for a general audience, but this one is now the best. The authors mix personal stories with Biblical insight and passionate argument to produce a book that is as creative as it is earnest and focused. This book is beautifully written and carefully argued. It would be the perfect book for a Bible study or church study group. Warning: it is an enjoyable read, but it might change your life."
   —Stephen H. Webb, Professor of Religion and Philosophy, Wabash
      College, author of *Good Eating* and *On God and Dogs: A Christian
      Theology of Compassion for Animals*

They Shall Not Hurt or Destroy: Animal Rights and Vegetarianism in the Western Religious Traditions by Vasu Murti, 2003, 137 pp.

"This is THE comprehensive reference work that documents the writings of mainstream religious and spiritual leaders who, from antiquity, have taught that compassion and kindness must mark our relationship with nonhuman as well as human beings. The author also reviews the observations of secular spokespersons, past and present, who through the centuries have championed the cause of compassion for animals, and for their rights as fellow beings."
   —Humane Religion

Omni-Science and the Human Destiny by Anthony Marr, 2003, 456 pp.

Wildlife preservationist Anthony Marr is no stranger to confrontation and danger. When he went to India for the third time to execute a 10-week tiger-saving expedition, he expected to fight poachers, illegal wood cutters, tiger bone traders, and smugglers. Unexpectedly, he encountered political corruption, organizational deceit, and personal betrayal that turned his world upside-down. This multi-faceted turmoil may have been responsible for the least expected encounter of all. The mysterious Raminothna who, deep in Tigerland, via a series of thoroughly logical steps, imparted upon him a new model of the Universe called Omniscientific Cosmology, which embraces all of the physical, biological, and social sciences, and shows the optimal human destiny and the fate of the Earth. Now, Anthony Marr must fight the battle of his life, one he must lose in order to win.

These books can be found at VegAdvPress.com.

Printed in the United States
127710LV00001B/3/P